HAZARDOUS AND INDUSTRIAL WASTES

Proceedings of the Thirty-Third Mid-Atlantic Industrial and Hazardous Waste Conference

EDITED BY
Nada Assaf-Anid
DEPARTMENT OF CHEMICAL ENGINEERING
MANHATTAN COLLEGE

TECHNOMIC PUBLISHING CO., INC.
LANCASTER • BASEL

First published 2001 by CRC Press
Taylor & Francis Group
6000 Broken Sound Parkway NW, Suite 300
Boca Raton, FL 33487-2742

Reissued 2018 by CRC Press

ISBN 13: 978-1-315-89373-0 (hbk)
ISBN 13: 978-1-351-07283-0 (ebk)

Visit the Taylor & Francis Web site at http://www.taylorandfrancis.com and the
CRC Press Web site at http://www.crcpress.com

HAZARDOUS AND INDUSTRIAL WASTES

Proceedings of the Thirty-Third Mid-Atlantic Industrial and Hazardous Waste Conference

June 18–20, 2001
Manhattan College
Riverdale NY, 10471

Universities and Centers participating in the Mid-Atlantic Industrial and Hazardous Waste Conference:

Auburn University
Center for Applied and Biocolloid Science, Australia
Chinese Academy of Science
City University of New York
Columbia University
University of Connecticut
Drexel University
Earth Engineering Center at Columbia University
Howard University
Indian Institute of Technology
University of Kentucky
Manhattan College
University of Maryland
University of Massachusetts
Massachusetts Institute of Technology
University of Missouri
The New Jersey Institute of Technology
University of Oklahoma
Purdue University
Rensselaer Polytechnic Institute
Rowan University
Rutgers, The State University of New Jersey
Smith College
State University of New York at Stony Brook
Texas A&M
Worcester Polytechnic

CONTENTS

---

*Not available at press time.

*Not available at press time.

*Not available at press time.

SYMPOSIUM 7:

Environmental Molecular Engineering

SYMPOSIUM 8:

Tracing and Geological Engineering

---

*Not available at press time.

*Not available at press time.

*Not available at press time.

PREFACE

The Mid-Atlantic Industrial and Hazardous Waste Conference is an annual meeting that brings together engineering and science professionals from academia, government and industry. At this year's meeting, more than seventy presentations and posters will present developments in research, regulation, and engineering practice. Topics include molecular engineering and computational chemistry; water quality modeling; novel treatment processes for industrial waste, sediments, groundwater, and gas-phase pollutants, including bioremediation and in-situ technologies; integrated waste management, composting and waste-to-energy technologies; environmental applications of colloidal and electrochemistry; sediment contamination; tracing and environmental geophysics, environmental geographical information systems (GIS); industrial ecology and sustainability; thermal and kinetic advances in pollution control and prevention; and environmental management and impact assessment. The conference includes a plenary session with presentations by the Donald O'Connor Chair of Environmental Engineering at Manhattan College, Dr. Dominic Di Toro; The Vinton Professor of Earth and Environmental Engineering at Columbia University, Dr. Peter Schlosser; and the Director of the Utah Water Research Laboratory at Utah State University, Dr. Ronald Sims.

NADA ASSAF-ANID
UPMANU LALL

ACKNOWLEDGEMENTS

Preparations for this conference started in September 2000 and have involved a large group of people. We gratefully acknowledge the generous assistance of the following contributors to this conference:

- Drs. Nickolas Themelis, Kevin Farley and Dominic DiToro, for their constructive advice and insight
- Ms. Joanne Carazo and the chemical engineering student helpers, for providing assistance with typing, mailing, printing, and communicating with presenters and chairs
- Mr. George Kuzma, for assisting with conference logistics
- Mr. Anthony Deraco from Technomic Publishing Co., Inc., for providing assistance with the preparation and publication of these *Proceedings*.

Special thanks go to all symposium chairs, paper and poster presenters, and all other participants who are the heart and pulse of the Conference. Please consult the remainder of this volume for the authors and their papers. Those interested in learning more about a particular topic should feel free to contact the authors directly.

The 33rd Mid-Atlantic Industrial & Hazardous Waste Conference gratefully acknowledges support from the following sponsors:

MANHATTAN COLLEGE · NEW YORK

EARTH ENGINEERING
Center
Columbia University

HydroQual

NEW YORK STATE LABORS
HEALTH AND SAFETY TRUST FUND

TECHNOMIC PUBLISHING CO.

SYMPOSIUM 1:

Sediment Contamination

IMPACTS OF A RARE FLOOD EVENT ON PCB FATE AND TRANSPORT IN THOMPSON ISLAND POOL, UPPER HUDSON RIVER

PETER H. ISRAELSSON
C. KIRK ZIEGLER
HARRY A. ZAHAKOS
JOHN P. CONNOLLY
Quantitative Environmental Analysis, LLC
305 West Grand Ave
Montvale, NJ 07645

INTRODUCTION

The sediment bed of the Upper Hudson River contains polychlorinated biphenyls (PCBs) as a result of historic discharges from General Electric (GE) plants located at Fort Edward and Hudson Falls, NY. Discharges from the GE plants ended in 1977. Since the mid-1970s, significant natural attenuation of surficial sediment PCB concentrations has occurred, primarily due to burial by cleaner sediment. The first reach of the river downstream of the GE plants is Thompson Island Pool (TIP), which is 9.6 km long and is terminated by a low-head dam (Thompson Island Dam), see Figure 1.

The average surficial (0-5 cm) PCB_{3+} (i.e., PCBs with 3 or more chlorine atoms) concentration in TIP cohesive bed areas has decreased from about 107 mg/kg (ppm) in 1977 to 14 ppm in 1998 [1]. However, elevated PCB concentrations still exist at depths generally greater than 10 cm, i.e., below the bioavailable zone, in cohesive bed areas. Thus, there is a concern that a rare flood event may cause sufficient erosion in TIP such that high PCB concentrations currently buried at depth would be reintroduced into the bioavailable zone.

A mathematical model was used to investigate the impacts of a rare flood on PCB fate in TIP. This comprehensive model was composed of three coupled sub-models: hydrodynamics, sediment transport and PCB_{3+} fate and transport. Results from extensive calibration and validation of the three sub-models showed that this modeling framework can be used with confidence to evaluate the impacts of a rare flood [1, 2, 3]. Note that this modeling framework was applied to the entire Upper Hudson River but, for convenience, this paper will focus on TIP results.

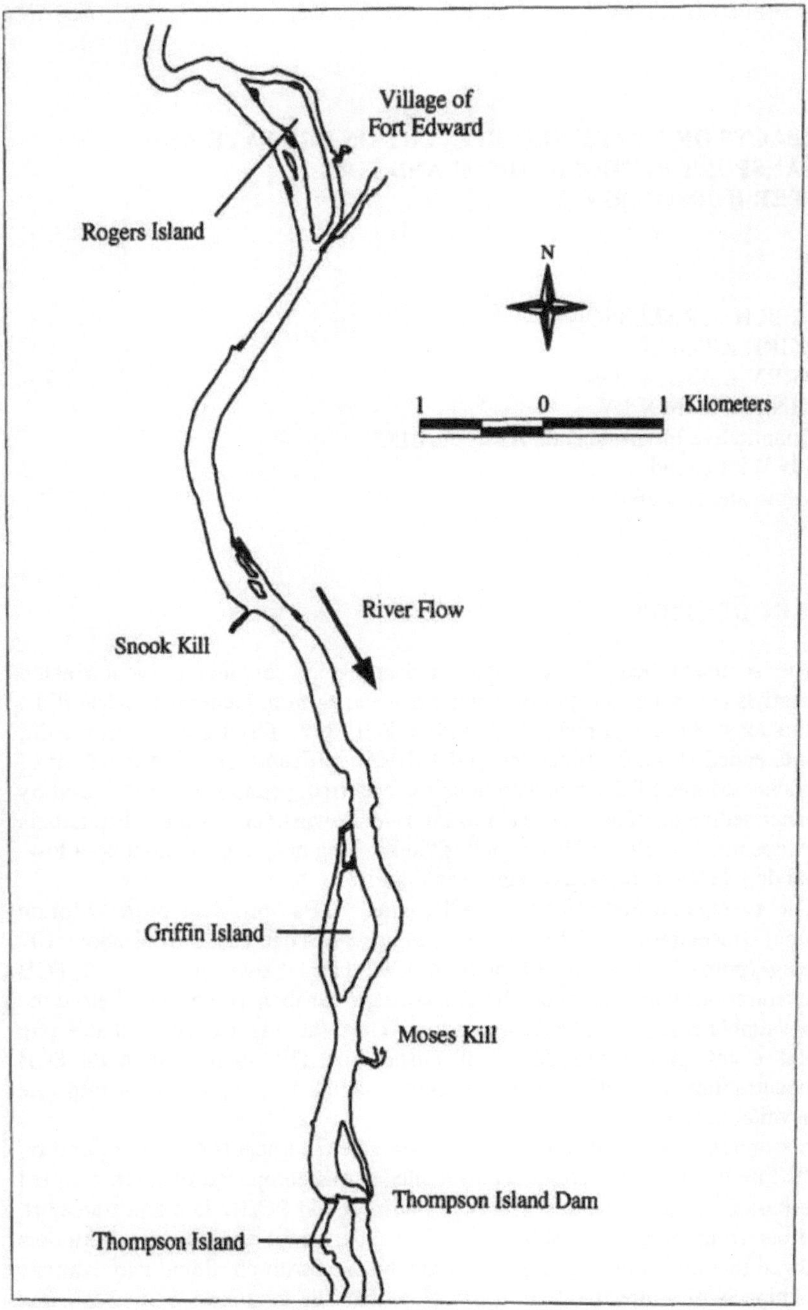

Figure 1. Thompson Island Pool, Upper Hudson River.

The primary objective of this paper is to present the results of a 100-year flood simulation in TIP and the impacts of this rare flood on PCB fate and transport in this reach of the Upper Hudson River. Additionally, this study should provide useful information to other modelers regarding methods for conducting a rare flood investigation. It should be noted that the information presented herein is a combination of results taken from the original Upper Hudson River model report [1] and the addendum to this report [3], which presents minor updates to the model.

MODEL DEVELOPMENT, CALIBRATION, AND VALIDATION

The development of a comprehensive mathematical model for determining the fate and transport of PCBs in TIP involved three steps: (1) collection and analysis of relevant environmental data, (2) selection and development of a model framework, and (3) calibration and validation of the model by comparing its results with historical data from TIP.

Site-specific data were the basis for assigning parameter values, either directly or by the process of model calibration. Each of the three sub-models was calibrated and validated using a data record that extends from 1977 to 2000. The extensive database makes TIP uniquely suited for the application of these models. Detailed descriptions of model development, calibration and validation are presented in QEA [1], Connolly et al. [2] and Ziegler et al. [3]. A brief review of the model structure and pertinent results will be presented here.

Two-dimensional, vertically-integrated hydrodynamic and sediment transport models were applied to TIP. The sediment transport model simulated cohesive and non-cohesive suspended load transport. A relatively high-resolution numerical grid was used, with 10 lateral and 68 longitudinal grid cells. Extensive site-specific data were used to develop model inputs and determine parameter values. Of particular importance to the rare flood simulation was the use of site-specific data for determining cohesive erosion parameters; no adjustment of cohesive resuspension parameters was made during model calibration. Results from the sediment transport model, i.e., resuspension and deposition fluxes, were transferred to the PCB fate model, which was a one-dimensional water-column model (with 12 longitudinal elements) that separately tracked PCB_{3+} concentration changes in the cohesive and non-cohesive portions of the sediment bed

The TIP sediment transport model was calibrated using total suspended solids (TSS) concentration data collected during the 1994 spring high flow (4-year flood) between March 31 to April 29, 1994. The calibration process involved comparing predicted and observed TSS concentrations at three locations in TIP. The model predicted net erosion from the non-cohesive sediment bed and net deposition in the cohesive sediment bed during this flood. Overall, the model computed net erosion of sediment from TIP during this 30-day period, with model results comparing well (18% error) with the estimate derived from a data-based mass balance analysis. Three simulations were conducted to validate the sediment transport model: (1) 1997 spring flood (~2-year flood); (2) 1993 spring flood (5-year flood); and (3) 23-year (May 1977 to July 2000) period [1, 3]. Note that two 10-year floods occurred during the 23-year simulation period.

The PCB fate model was calibrated and validated using sediment and water column PCB data sets collected at the site since the late 1970s. The calibration period was from the time of the first extensive sediment sampling study in 1977 to the 1998 sediment survey. Model sediment calibration was performed by comparing results to average TIP PCB_{3+} concentrations in cohesive and non-cohesive sediments. Validation was accomplished by extending the simulation to 2000 and comparing results to extensive water column data collected from 1997 to 2000. In addition, predicted sediment PCB_{3+} concentrations were compared to 1991 and 1998 sediment data on a finer spatial resolution [1, 2, 3]. The ability of the model to accurately predict water column PCB_{3+} concentrations, coupled with the accurate prediction of surface sediment concentration changes between 1977 and 2000, indicates that it properly represents the movement of PCBs between the sediment and the water column.

DEVELOPMENT OF MODEL INPUTS FOR A RARE FLOOD

The impacts of a 100-year flood on sediment transport and PCB fate processes in TIP were investigated using the calibrated and validated models. No model parameters were adjusted during this simulation. All parameters were set at the values determined during calibration. Boundary conditions and sediment initial condition specifications are discussed below.

The peak flow rate for a 100-year flood at Fort Edward has been estimated to be 1,339 m^3/s (47,330 cfs), on a daily-average basis [4]. Flood hydrographs in TIP are variable, making it necessary to construct upstream and tributary hydrographs for the 100-year flood that are representative of average flood conditions. A normalized 8-day flood hydrograph was developed from nine floods (with peak daily average flow rates ranging from 606 to 965 m^3/s) that occurred occurred between 1977 and 1997 [1]. The 100-year flood peak flow rate was applied to the normalized hydrograph and used to specify the upstream boundary condition.

Peak daily-average flow rates in TIP tributaries (Snook Kill, Moses Kill and direct runoff) during the 100-year flood were estimated using results of flood insurance studies conducted by Federal Emergency Management Agency (FEMA). The FEMA studies indicated a daily-average runoff rate from the drainage basin during a 100-year flood on the Upper Hudson River of about 0.20 $m^3/s\text{-}km^2$ [5]. This runoff rate was applied to all tributaries on the day of peak flow at Fort Edward; tributary flow rates for this day were specified by multiplying the runoff rate by the tributary drainage area. Model input flows for the other flood days were calculated by scaling the normalized flood hydrograph developed for each tributary to the tributary flow rate on the day of peak flow at Fort Edward [1].

Sediment loads from upstream and tributary sources were specified using sediment rating curves developed from TSS concentration data collected at Fort Edward, Snook Kill and Moses Kill [1, 3]. Total sediment loading to TIP during the 100-year flood was about 69,000 tons. This load corresponds to approximately 198% of the estimated average annual TIP sediment load during the long-term simulation of about 34,800 tons/year.

Erosion in the non-cohesive areas is dependent on the bed properties (e.g., d_{50}, fractions of coarse and fine material), which vary over time. The sensitivity of the model results to non-cohesive bed property distributions was investigated by comparing a series of 100-year flood simulations using bed property distributions calculated after each year of the long-term simulation [1]. The results indicated that the bed property variations had little impact on 100-year flood results. The distribution calculated at the end of 1987 was used in the results presented below because it was closest to the average of the other simulations.

The results of the hydrodynamic and sediment transport model were used to determine flows, volumes, depths, TSS concentrations, and deposition and resuspension rates in the PCB fate model in the same manner as for the model calibration. The flood was assumed to occur in 2000 immediately after the calibration period; sediment PCB_{3+} concentrations at this time were used as initial conditions for the flood. PCB_{3+} loading at the upstream and tributary inflow boundaries were assumed to be zero during the flood in order to assess the PCB_{3+} load due only to TIP sediment erosion.

RARE FLOOD SIMULATION RESULTS

The sediment transport model predicted mean erosion depths of 0.79 and 0.12 cm for the cohesive and non-cohesive bed in TIP, respectively [3]. Maximum erosion depths for both the cohesive and non-cohesive bed were approximately 8 cm. Erosion occurred in approximately 93% of the cohesive bed area, with net deposition in the remaining area. Erosion depths of 2 cm or less were predicted for 85% of the cohesive bed area and about 3% of the cohesive bed area had erosion depths greater than 5 cm (Figure 2). Scour depths of 1 cm or less were predicted in approximately 98% of the non-cohesive bed area (Figure 2).

The model predicted that a total of 4,800 tons of sediment would be eroded and transported out of TIP during the 100-year flood, which corresponds to about 7% of the total sediment load input. Of this amount, 2,700 and 2,100 tons of net erosion occurred in the cohesive and non-cohesive bed areas of TIP, respectively. For comparison, net deposition was predicted in the cohesive bed areas during the 1993 and 1994 spring floods (1,360 and 990 tons of net deposition, respectively). In the non-cohesive bed, 1,330 and 1,520 tons of net erosion were predicted during the 1993 and 1994 floods, respectively [1, 3].

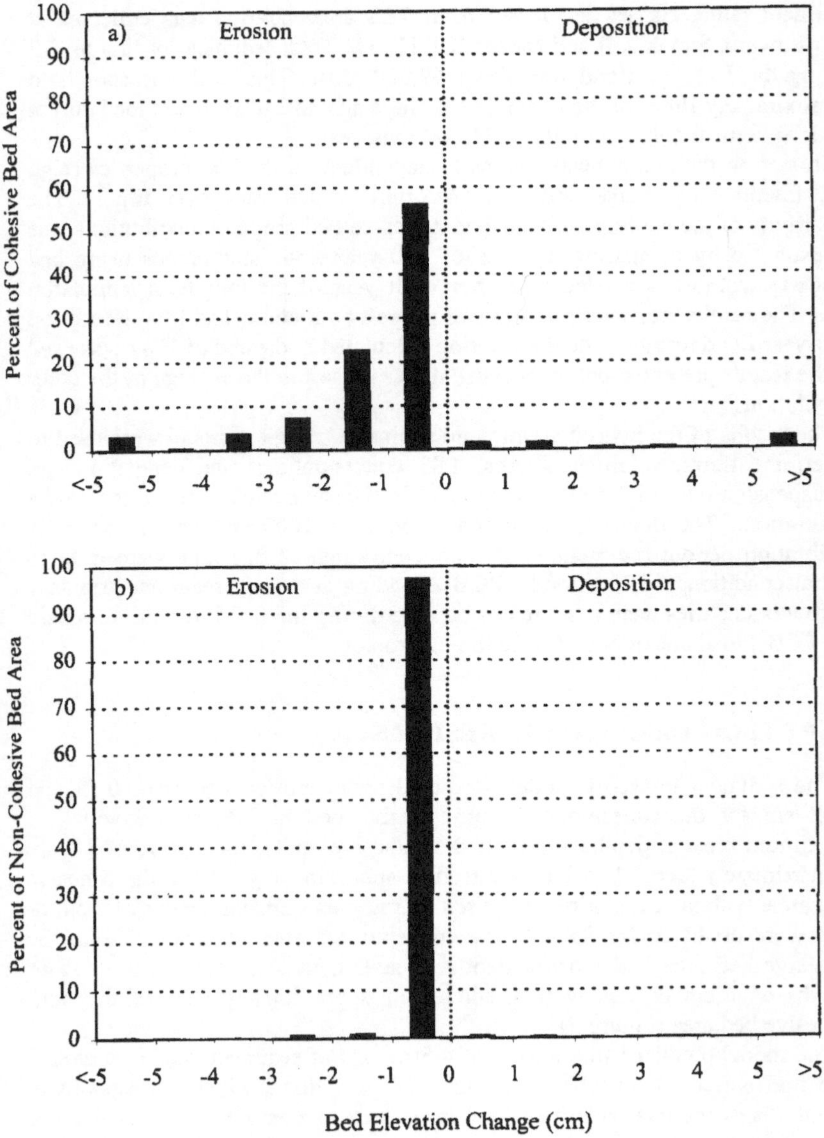

Figure 2. Predicted distributions of (a) cohesive and (b) non-cohesive bed elevation changes in TIP at the end of the 100-year flood.

The model predicted that 53 kg of PCB_{3+} would be eroded from TIP sediments. Of the mass eroded, 97% were predicted to be transported out of TIP to the downstream reaches of the Upper Hudson River [3]. The 1997 USEPA Data Evaluation and Interpretation Report provided estimates of the 1984 PCB_{3+} mass inventory in TIP, the lowest estimate of which was 14,500 kg [6]. Assuming comparable PCB_{3+} mass inventory in 2000, the 100-year flood would only erode about 0.4% of the total TIP PCB_{3+} inventory.

The impact of a 100-year flood on long-term PCB fate in TIP is illustrated by comparing results of 30-year forecast simulations with and without this flood that were conducted with the original model [1]. The first forecast simulation extended from 2000 to 2030 with no 100-year flood. The second forecast simulation was identical to the first simulation except for the inclusion of a 100-year flood in 2006. The occurrence of a 100-year flood in 2006 was predicted to increase the average TIP cohesive surface sediment PCB_{3+} concentration from 4.3 ppm to 5.8 ppm. The predicted increases would effectively set back the recovery in these sediments by about 4 years, bringing concentrations to levels predicted in 2002. Thus, the 100-year flood has minimal impact on long-term natural recovery of PCB bed concentrations in TIP cohesive deposits.

VALIDITY OF 100-YEAR FLOOD SIMULATION

The 100-year flood simulation represents an extrapolation beyond the range of TIP conditions included in sediment transport model calibration and validation simulations, which creates uncertainty in the 100-year results and brings into question the accuracy of the predictions. The validity of these predictions was investigated by comparing 100-year flood results to calibration and validation results in order to evaluate the significance of extrapolating beyond data-based simulations.

An important validation of the sediment transport and PCB fate models, with respect to evaluating the impacts of a rare flood, was accomplished through simulation of a large flood that occurred in TIP during early January 1998 [1]. The peak of this flood was on January 10, with a Fort Edward daily average flow rate of 957 m^3/s, which approximately corresponds to a 10-year flood. Since 1977, the January 1998 and May 1983 floods are the largest observed at Fort Edward, with the peak daily average flow during the May 1983 flood being 965 m^3/s. Water column PCB and TSS concentration data collected at Fort Edward and Schuylerville (about 11 km south of TIP) during the January 1998 flood indicate that sediment bed erosion occurred between those two locations. Without adjustment of model parameters, the model was able to predict the observed increase in water column PCB_{3+} load between Fort Edward and Schuylerville with excellent accuracy. The increase in PCB_{3+} load between these two locations during this flood was caused entirely by sediment bed erosion. A data-based mass balance indicated that 30 kg of PCB_{3+} were eroded from the bed between Fort Edward and Schulerville during this flood;

the model predicted loading increase of 40 kg compares very well with observed value [7]. Thus, these results show that the sediment transport model is capable of accurately predicting sediment bed erosion in TIP (and other portions of the Upper Hudson River) during an extreme event.

The extent to which the 100-year flood represents an extrapolation of calibration and validation conditions can be evaluated by comparing peak flow rates. The 1993 (821 m^3/s) and 1994 (784 m^3/s) spring floods, which were used for sediment transport model calibration and validation, had peak flows that were about 60% of the 100-year peak flow. Similarly, the 10-year floods that occurred in 1983 (965 m^3/s) and 1998 (957 m^3/s), both of which were included in the long-term validation simulation, had peak discharge that were about 28% lower than the 100-year flood. Furthermore, the predicted maximum velocities during the 100-year flood were typically about 45% and 30% higher than those predicted during 1993-94 and 1998 floods, respectively. Of particular importance to sediment transport processes is bottom shear stress; the median value of maximum predicted bottom shear stress during the 100-year flood was about 100% and 75% higher than during the 1993-94 and 1998 floods, respectively [1].

The above results strongly indicate that the 100-year flood, while certainly an extrapolation beyond calibration and validation simulations, is not an extreme extrapolation. Therefore, the 100-year flood results can be considered a realistic and valid extrapolation beyond calibration and validation conditions.

RESULTS FROM OTHER TIP MODELING STUDIES

USEPA has also conducted a 100-year flood simulation for TIP using a two-dimensional, vertically-averaged hydrodynamic model [8]. Although the analysis did not use a calibrated sediment transport model, the results indicated that the average scour depth in cohesive bed areas of TIP is 0.3 to 0.4 cm [8], which is about a factor of two lower than the predictions presented above.

A one-dimensional non-cohesive sediment transport model (HEC-6) was applied to TIP by Zimmie [9]. While the model predictions are uncertain because of simplifying assumptions and lack of rigorous calibration, comparisons of the results to those presented above are informative. Zimmie [9] simulated a flood with a peak flow rate within 2% of the 100-year flood described above. The predicted average and maximum erosion depths in non-cohesive areas were 0.5 and 5 cm, respectively [9]. While these results indicate a slightly greater average non-cohesive erosional depth than the present model, the predictions of all three TIP models yield the same basic conclusion: the 100-year flood causes relatively minor erosion in TIP.

SENSITIVITY ANALYSIS

Sensitivity of the 100-year flood simulation in TIP to input sediment transport parameters was investigated. Five parameters were varied within reasonable ranges [1]: (1) cohesive resuspension constant, a_o; (2) sand fraction of incoming sediment load, f_2; (3) effective particle diameter of coarse (sand) class sediment, d_2; (4) non-cohesive bottom roughness, z_o; and (5) non-cohesive bed armoring constant, B. The model was most sensitive to variations in B, with a maximum increase in non-cohesive bed erosion of 82%. Maximum variations in cohesive sediment erosion due to parameter variation were lower.

Sensitivity to sediment loading and hydrographs were also investigated using a zero sediment loading assumption and 95% confidence intervals about the hydrograph [1]. Variations had a minimal impact on erosion depths and total mass eroded.

Finally, the impact of including flood plains in the model calculations was also evaluated by using a modified grid [1]. The inclusion of flood plains caused a decrease in net sediment erosion as well as a reduction in average and maximum channel scour depths. Thus, the original 100-year flood calculation, which neglected flood plain effects, produced conservative results.

CONCLUSIONS

The 100-year flood results indicated that relatively minor erosion, in both cohesive and non-cohesive bed areas, will occur in TIP, with mean erosion depths of less than 1 cm and maximum scour depths of about 8 cm. The model predicted that approximately 53 kg of PCB_{3+} would be eroded during the flood period, which corresponds to approximately 0.4% of the USEPA estimate of 1984 sediment PCB_{3+} mass. Thus, the analysis indicates that a 100-year flood would not have a major impact on the long-term natural recovery of the sediments. The results for this flood were of the same order of magnitude as predictions made by USEPA [8] and Zimmie [9] for a similar high flow event. Comparison of the 100-year flood to historical flood results, i.e., 1993, 1994, 1997 and 1998 floods, showed that simulation of this rare flood event is not an extreme extrapolation from the calibration and validation simulations. Sensitivity analyses indicated that uncertainty in the 100-year flood projections due to parameter and boundary condition variation was less than a factor of two.

ACKNOWLEDGEMENTS

This work was supported by the General Electric Company. It was managed by John Haggard and Bob Gibson, who provided both review and technical guidance.

REFERENCES

1. Quantitative Environmental Analysis. 1999. "PCBs in the Upper Hudson River," report prepared for the General Electric Company, Albany, NY.

2. Connolly, J.P., H.A. Zahakos, J. Benaman, C.K. Ziegler, J.R. Rhea, and K. Russell. 2000. "A model of PCB fate in the Upper Hudson River," *Environ. Sci. Technol.*, 34(19):4076-4087.

3. Ziegler, C.K., P.H. Israelsson, H.A. Zahakos, D. Glaser, and J.P. Connolly. 2001. "Addendum to PCBs in the Upper Hudson River July 1999," report prepared for the General Electric Company, Albany, NY.

4. U.S. Environmental Protection Agency. 1996. "Phase 2 Report – Review copy: Further Site Characterization and Analysis – Volume 2B Preliminary Model Calibration Report." Hudson River PCBs Reassessment RI/FS. Volume 2B, Book 1 of 2. USEPA, Region 2, New York, New York.

5. Federal Emergency Management Agency. 1995. "Flood Insurance Study, Saratoga County, New York (All Jurisdictions)." August 16, 1995.

6. U.S. Environmental Protection Agency. 1997. "Phase 2 Report – Review Copy: Further Site Characterization and Analysis - Volume 2C Data Evaluation and Interpretation Report." Hudson River PCBs Reassessment RI/FS. USEPA, Region 2. New York, New York.

7. Ziegler, C.K., P.H. Israelsson, and J.P. Connolly. 2001. "Stability of Soft Sediment Deposits in the Upper Hudson River," report prepared for the General Electric Company, Albany, NY.

8. U.S. Environmental Protection Agency. 2000. "Phase 2 Report – Review copy: Further Site Characterization and Analysis – Volume 2D Revised Baseline Modeling Report." Hudson River PCBs Reassessment RI/FS. Volume 2D, Book 1 of 4. USEPA, Region 2, New York, New York.

9. Zimmie, T.F. 1985. "Assessment of Erodibility of Sediments in the Thompson Island Pool of the Hudson River," Final Report, NYSDEC, RPI Project No. 5-25360.

Cohesive Sediment Resuspension: Experimentation and Analysis

Parmeshwar L. Shrestha, Imali D. Kaluarachchi, Paul J. Anid,
Alan F. Blumberg, and Dominic M. DiToro
HydroQual Inc., One Lethbridge Plaza, Mahwah, NJ 07430
PH (201) 529-5151; FAX (201) 529-5728; email: pshrestha@hydroqual.com

Abstract

The resuspension characteristics of cohesive bed sediment samples were assessed through
laboratory experiments using a device called the "shaker." The samples were reconstructed in the laboratory from sediments obtained from ocean, estuarine, and riverine beds. Each sample was subjected to a series of increasing shear stresses, with each shear stress being maintained for a certain period. For each experiment, it was found that after a certain time into the resuspension process, the turbidity in the overlying water column reached a steady state indicating that net erosion at that time was equal to zero. The experimental observations affirm the theory that the bed armors after a finite amount of sediment has been resuspended into the water column due to the presence of underlying consolidated layers of greater shear strength. This prevents the underlying layers and particles from being exposed to the flow. At this point, the bed becomes armored. The steady state concentrations as well as the pre-steady state sediment concentrations in the overlying water column are influenced by the characteristics of the sediment bed and notably by the consolidation period of the sediment layers. For increasing bed shear stress, a longer time is required to attain steady-state concentrations in the water column. This is due to the variation in sediment strength with depth of the bed.

AN INDUSTRIAL ECOLOGY APPROACH TO DREDGING THE NEW YORK/NEW JERSEY HARBOR: THE CASE OF MERCURY

Thomas Wakeman[1] and Nickolas J. Themelis[2]

1. General Manager, Waterways Development, Port Department, Port Authority of New York and New Jersey, One World Trade Center, 62W, New York, NY 10048 (twakeman@panynj.gov).
2. Director, Earth Engineering Center, Columbia University, 500 West 120th St., Mudd Hall, Room 1047, New York, NY, 10027 USA (njt1@columbia.edu)

Abstract

Several million cubic meters of sediments must be dredged annually to clear navigation channels, berths, and marinas in New York/new Jersey Harbor. An estimated 300 million tons have been dredged and discharged during the period of 1900-2000. In the past, the dredged material was disposed in a designated ocean site in the New York Bight. However, new sediment testing procedures were implemented towards the end of the 20th century and much of the harbor's dredged material was classified as unsuitable for ocean placement. Ironically, these restrictions came at a time when the quality of harbor's sediments is improving as a result of pollution controls imposed by the Clean Water Act and increased public awareness of the need to curb pollution of river and coastal waters. For example, with the exception of a few areas where historical contamination continues, the harbor-wide concentration of mercury has decreased to 0.7-0.8 parts per million, a level that is approaching the pre-industrial background level. Anthropogenic contamination of the harbor's sediments has occurred for centuries. Contaminants from land-based sources introduced into the water either as soluble or particulate matter over the 42,000 square kilometers of the Hudson-Raritan basin end up associated with particles that settle to the bottom, primarily in artificially deepen areas such as berths and navigation channels. Development of a basin-wide and site-specific sediment management strategy is necessary to guide the continuing clean-up of new and historical contamination and the dredging program that is the lifeline of the major port on the East coast. The backbone of this strategy is the integration of data from a comprehensive field monitoring and modeling program with a parallel investigation of watershed and airshed sources and sinks using industrial ecology methodology. As an example, this paper will concentrate on one of the principal sediment contaminants: Mercury. It examines the results of a study on new and historical sources of mercury accumulation in the Harbor and presents preliminary estimates of the material balance between inputs and outputs.

SYMPOSIUM 2:

Colloids and Electrochemical Methods

The 33rd Mid-Atlantic Industrial & Hazardous Waste Conference
Manhattan College
Riverdale, NY
June 18 – 20, 2001

TITLE: Electrochemical Coagulation for Oily Wastewater Treatment

Chen-Lu Yang
Technical Consultant
43 Rector Street, Millburn, NJ 07041

E-MAIL: yang1987@aol.com

ABSTRACT: Motor oil and industrial detergent were used to prepare simulated oily wastewater. When stabilized with surfactant, the turbidity (Formazin Attenuation Unit, FAU) was found to have a positive correlation with oil content in the emulsion with a correlation coefficient of 2.78 FAU/(mg/L). The oil/water emulsion was tested with filtration, decantation, centrifugation, flotation, and electrochemical coagulation for the effectiveness of each process on demulsification. Electrochemical coagulation was selected for further study. DC voltage was applied to the electrodes, dissolving ferrous ions at the anode and forming hydrogen and hydroxyl ions at the cathode. A few hundred parts per million of sodium chloride was added to the solution to provide ionic conductivity and to prevent passivation of the iron electrode. The ferrous ion was oxidized to ferric ion and that destabilized the emulsion. The coalesced oil droplets adsorbed onto the highly dispersed and reactive ferric hydroxide coagulant. The oil-rich sludge that was generated in the operation was floated to the surface where it formed a blanket that was removed by skimming. The complex mechanism was completed in four minutes. When the reactor was operated at four minutes, the dose of ferric ions generated in the process was 165 mg/L. The treatment reduced the turbidity of the emulsion from 1700 FAU to 60 FAU. The addition of more iron to the system beyond the four-minute treatment period resulted in further change in the turbidity in a much slow rate. In a continuous operation at a current of 2 amperes and a throughput of 320 mL/min, the turbidity of the electrically treated effluent was less than 14 FAU that is the detection limit of the DR/4000 Spectrophotometer.

Nonylphenol and Its Ethoxylates in Wastewater Treatment Plants

33rd Mid-Atlantic Hazardous and Industrial Wastes Conference,
Manhattan College, Riverdale, NY

Kauser Jahan*,Ph.D., P.E., Associate Professor, Department of Civil & Environmental Engineering,
Rowan University, 201 Mullica Hill Road, Glassboro, NJ 08028 U.S.A.
*** Corresponding Author**
Telephone: 856-256-5323
Fax: 856-256-5242
email: jahan@rowan.edu

Maragaret Jacques, Research Assistant, Rowan University
Shira Perlis, Research Assistant, Rowan University
Marcus Roorda, Research Assistant, Rowan University

ABSTRACT FOR ORAL PRESENTATION

Certain non-ionic surfactants, more specifically alkylphenolethoxylates (APEs), have been shown to be estrogenic and toxic to aquatic organisms according to recent studies conducted primarily in Europe, Japan and Canada. Recent evidence that some APE breakdown products are **estrogenic** has intensified concern over their environmental and human health effects. APEs are nonionic surfactants made up of a branched chain ethylene oxide to produce an ethoxylate chain. The main alkylphenols used are nonylphenol (NP) and octylphenol (OP). Nonylphenol ethoxylates (NPEs) encompass about 80% of the world market, and octylphenol ethoxylates (OPEs) represent most of the rest. They have been used extensively for their effectiveness, economy and ease of handling and formulating for more than forty years. Most APEs enter the aquatic environment from wastewater treatment plant discharges. In the United States, industrial uses of APE encompass the largest category (55%). There are no U.S. regulatory actions to date for APEs. Therefore, more research on the fate and transport of APEs in the environment is essential. This paper will focus on attempts to characterize the occurrence of nonylphenol and its ethoxylates in wastewater treatment plants. Studies are currently underway to determine the presence of these compounds in local wastewater treatment plants. This is important as the state of New Jersey has a strong industrial base comprising of pharmaceutical and chemical manufacturers, which are the major users and generators of APEs.

PHYSICAL MECHANISMS OF COLLOIDAL SILICA GROUTING IN REDMEDIATION OF CHROMIUM CONTAMINATED SOIL

NETNAPID YOSSAPOL
Civil and Environmental Department
New Jersey Institute of Technology
University Heights,
Newark, NJ 07102

JAY NAMUNU MEEGODA
Civil and Environmental Department
New Jersey Institute of Technology
University Heights,
Newark, NJ 07102

INTRODUCTION

In this paper, the mechanisms of colloidal silica grouting during stabilization of chromium contaminated soil are explained by using a mathematical model. Grouting is a technique to inject low viscous time-setting material into media that needs increased strength or improved engineering performance. The material injected has the ability to penetrate into void spaces in soils and rock, later hardening inside the media. This causes an increasing in the mechanical strength of the media and reduces the permeability of soil preventing chemical from migration [1]. This technique is used in environmental application by injecting stabilizer into contaminated soil media. This provides in situ mixing of the grout with the contaminated soil, preventing the further contamination. In situ S/S techniques are usually less labor and energy intensive than ex situ techniques that require excavation, transport and disposal of the treated material [2].

Grout can be chosen from variety of products depending on soil and site conditions. Several studies have been accomplished in applying cement-base grouts as a waste barrier. Allan and Kukacka 1994 [3] showed that super-plasticizers created an effective barrier to control chromium migration into fine-grained soils. This technology was then successfully applied to an in situ Mixed Waste Landfill Site at Sandia National Laboratory, New Mexico.

Colloidal silica (CS) is an alternative grout. It can increase strength and reduce permeability of soil. Base on the long-term tests, Watana et al. 1987 [4], and Persoff et al. 1999 [5] demonstrated that CS has excellent grout properties such as high strength, low syneresis and low permeability. CS has advantages over other grouts in terms of low viscosity, which can penetrate into void spaces between fine-grained soil [6]. In addition, CS has large surface areas that increase the possibility of adsorbing metal contaminants onto its surface. Noll et al. 1992 [1] first proposed the application of colloidal silica grouting for the remediation of contaminated soils Persoff et al. 1995 [7] and Apps et al. 1998 [8] performed a series of laboratory experiments where CS grouting behavior was observed and gellation was simulated. Morisdis et al. 1997 [9] injected CS into soil creating a barrier wall, preventing the further migration of radionuclides. Persoff et al. 1998 [10] suggested test methods for CS application as waste barrier for site specific conditions.

The in-situ S/S of metal contaminated soils using CS is a well established technology. However, the theoretical understanding of how the CS function as an effective S/S agent is lacking. This research attempts to close that gap by performing a theoretical study to understand the stabilization mechanisms of CS grouts. In this paper, the mechanisms of colloidal silica grouting on soil permeability were explained by microscopic model of fluid flow through soil.

THEORY

Soil permeability is a macroscopic parameter governed by the movement of pore fluid through soil at the microscopic level. At the microscopic level, the steady state flow of an incompressible fluid in an incompressible porous media is regulated by the equation of continuity and steady state Navier-Stokes equation. In this research, the microstructure of CS treated contaminated soils is obtained by solving the Navier-Stokes equation with appropriates boundary conditions. The assumption is that porous soil mass consists of spatially periodic spherical particles of a given diameter. Thus, it is possible to analyze a unit representative cell consisting of one particle with its surrounding ambient space. Based on this assumption, the influence of shape, non-periodic nature, and size can be calculated by alternation of the unit cell. This will be lead to the derivation of a governing equation for the flow through soil in terms of macroscopic parameters, which can be physically measured.

The flow field through the spatially periodic porous medium for a unit cell shown in Figure 1. can be written using the steady-state Navier-Strokes equation:

$$\rho \underline{V} \cdot \nabla \underline{V} = -\nabla p + \mu \nabla^2 \underline{V} \tag{1}$$

and for continuity of fluid motion for an incompressible fluid:

$$\nabla \cdot \underline{V} = 0 \tag{2}$$

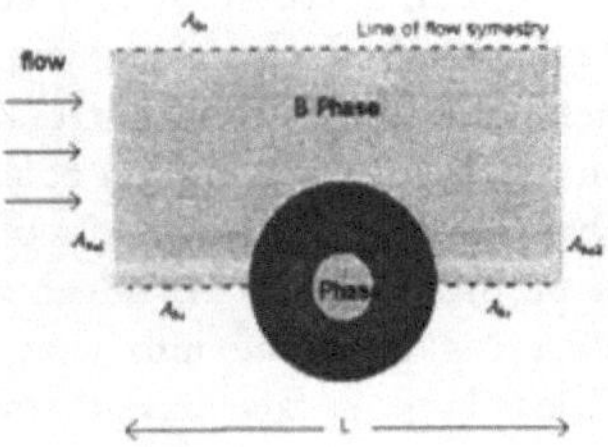

Figure 1. Unit Cell of Porous Media

boundary conditions on velocity:

no-slip: $\underline{V} = 0$ on A_{ab}

symmetry: $\underline{n} \cdot \nabla \underline{V} = 0$ on A_{bs}

periodicity: $\underline{V}(\underline{r} + n_i \underline{l}_i) = \underline{V}(\underline{r})$ on A_{be}

boundary condition on pressure:

periodicity: $P(\underline{r} + n_i \underline{l}_i) = P(\underline{r}) + n_i \underline{l}_i \Delta P$

Where: ΔP = pressure drop across the inlet and exit boundaries

A_{ba}, A_{bs}, A_{be} = boundary of unit cell as shown in Figure 1.

ρ = mass density

μ = coefficient of absolute viscosity

The periodic boundary conditions are defined using the translation vector given by the following. If the porous medium is represented as a repetitive array of rigid particles extending indefinitely in all directions, the location of any lattice point is given by:

$$\underline{R} = \underline{R}_o + n_i \underline{l}_i \quad i = 1,2,3,... \tag{3}$$

Where: $\underline{R}_0$ = some reference position

n_i = integers

$\underline{l}_i$ = non-unique set of basis vectors called the lattice vectors

The volume of the array is partitioned into unit cells, defined such that the entire medium can be produced from a single cell by the repeated use of the translation given by Equation (3).

Numerical Implementation

A triangular element was used for the finite element implementation of the above. The two horizontal velocities U and V (perpendicular to each other) were allowed to vary as quadratic functions over the element whereas pressure was allowed to vary linearly.

The Navier-Stokes equation for two-dimensional case can be rewritten as:
x direction or u momentum:

$$\rho\left[U\frac{\partial U}{\partial x} + V\frac{\partial U}{\partial y}\right] = -\frac{\partial p}{\partial x} + \varepsilon_x\left[\frac{\partial^2 U}{\partial x^2} + \frac{\partial^2 U}{\partial y^2}\right] \tag{4}$$

y direction or v momentum:

$$\rho\left[U\frac{\partial V}{\partial x} + V\frac{\partial V}{\partial y}\right] = -\frac{\partial p}{\partial y} + \varepsilon_y\left[\frac{\partial^2 V}{\partial x^2} + \frac{\partial^2 V}{\partial y^2}\right] \tag{5}$$

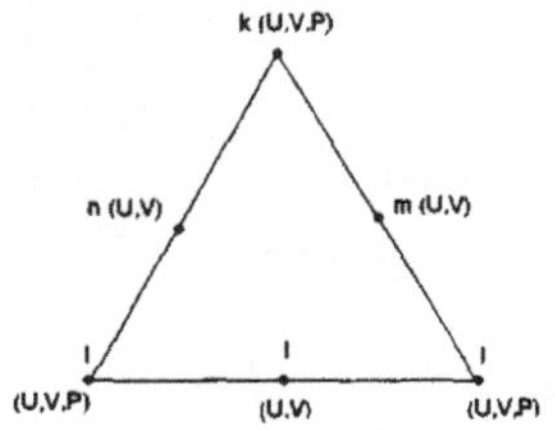

Figure 2. A Triangle for Finite Element

and the continuity equation as:

$$\frac{\partial U}{\partial x}+\frac{\partial V}{\partial y}=0 \tag{6}$$

Where: $\varepsilon_x = \varepsilon_y = \mu$

When the Galerkin approach is utilized, the governing equations may be stated in terms of two basis function weighting factors as:

u momentum:

$$\{f_1^i\}=\int_V \langle N_i\rangle^T \left\{\rho\left[U\frac{\partial U}{\partial x}+V\frac{\partial U}{\partial y}\right]-\varepsilon_x\left[\frac{\partial^2 U}{\partial x^2}+\frac{\partial^2 U}{\partial y^2}\right]+\frac{\partial p}{\partial x}\right\}dV=0 \tag{7}$$

v momentum:

$$\{f_1^i\}=\int_V \langle N_i\rangle^T \left\{\rho\left[U\frac{\partial V}{\partial x}+V\frac{\partial V}{\partial y}\right]-\varepsilon_y\left[\frac{\partial^2 V}{\partial x^2}+\frac{\partial^2 V}{\partial y^2}\right]+\frac{\partial p}{\partial y}\right\}dV=0 \tag{8}$$

continuity:

$$\{f_1^i\}=\int_V \langle M_i\rangle^T \left\{\frac{\partial U}{\partial x}+\frac{\partial V}{\partial y}\right\}dV=0 \tag{9}$$

Where:

$\langle N_i\rangle$ is the quadratic basis function defined over all elements adjacent to node i

or $N_i(x,y)=1$ at node i and $N_i(x,y)=0$ at all other nodes

$\langle M_i\rangle$ is the linear basis function defined over all elements adjacent to node i

or $M_i(x,y)=1$ at node i and $M_i(x,y)=0$ at all other corner nodes

Equations (7) and (8) contain second order derivatives, which cannot be handled directly. Therefore, to lower the order of the derivative, it is necessary to use Green's transformation:

$$\{f_1^i\}=\int_V \left\{\langle N_i\rangle^T\left[\rho\left(U\frac{\partial U}{\partial x}+V\frac{\partial U}{\partial y}\right)\right]-\left[\varepsilon_x\left(\left\langle\frac{\partial N_i}{\partial x}\right\rangle^T\frac{\partial U}{\partial x}+\left\langle\frac{\partial N_i}{\partial y}\right\rangle^T\frac{\partial U}{\partial y}\right)\right]-\left\langle\frac{\partial N_i}{\partial x}\right\rangle^T p\right\}dV$$
$$-\int_S \left\{\varepsilon_x\langle N_i\rangle^T\left[\frac{\partial U}{\partial x}C_x+\frac{\partial U}{\partial y}C_y\right]-\langle N_i\rangle^T pC_x\right\}dS=0 \tag{10}$$

$$\{f_2^i\}=\int_V \left\{\langle N_i\rangle^T\left[\rho\left(U\frac{\partial V}{\partial x}+V\frac{\partial V}{\partial y}\right)\right]-\left[\varepsilon_y\left(\left\langle\frac{\partial N_i}{\partial x}\right\rangle^T\frac{\partial V}{\partial x}+\left\langle\frac{\partial N_i}{\partial y}\right\rangle^T\frac{\partial V}{\partial y}\right)\right]-\left\langle\frac{\partial N_i}{\partial y}\right\rangle^T p\right\}dV$$
$$-\int_S \left\{\varepsilon_y\langle N_i\rangle^T\left[\frac{\partial V}{\partial x}C_x+\frac{\partial V}{\partial y}C_y\right]-\langle N_i\rangle^T pC_y\right\}dS=0 \tag{11}$$

where C_x, C_y = direction cosines normal to the surface S

and
$$\{f_1^i\}=\int_V\langle M_i\rangle^T\left\{\frac{\partial U}{\partial x}+\frac{\partial V}{\partial y}\right\}dV=0 \tag{12}$$

The surface integrals of Equations (10) and (11) may be neglected for inter-element interfaces but must be accounted for on the boundary of the system. These integrals may be computed from the known boundary conditions and incorporated into the right hand side vector of the final set of simultaneous equations.

When the Newton-Raphson method is used, the terms of the element coefficient matrix may be written in the form shown below:

$$\left.\begin{array}{l}\frac{\partial f_1^i}{\partial U}\Delta U+\frac{\partial f_1^i}{\partial V}\Delta V+\frac{\partial f_1^i}{\partial p}\Delta p=\Delta f_1\\ \frac{\partial f_2^i}{\partial U}\Delta U+\frac{\partial f_2^i}{\partial V}\Delta V+\frac{\partial f_2^i}{\partial p}\Delta p=\Delta f_2\\ \frac{\partial f_3^i}{\partial U}\Delta U+\frac{\partial f_3^i}{\partial V}\Delta V+\frac{\partial f_3^i}{\partial p}\Delta p=\Delta f_3\end{array}\right\}\qquad \begin{array}{c}\\ 6rows\\ 6rows\\ 6rows\end{array}\overset{\begin{array}{ccc}6cols & 6cols & 6cols\end{array}}{\begin{bmatrix}\frac{\partial f_1}{\partial U} & \frac{\partial f_1}{\partial V} & \frac{\partial f_1}{\partial p}\\ \frac{\partial f_2}{\partial U} & \frac{\partial f_2}{\partial V} & \frac{\partial f_2}{\partial V}\\ \frac{\partial f_3}{\partial U} & \frac{\partial f_3}{\partial U} & \frac{\partial f_3}{\partial U}\end{bmatrix}}\begin{bmatrix}\Delta U\\ \Delta V\\ \Delta p\end{bmatrix}=\begin{bmatrix}\Delta f_1\\ \Delta f_2\\ \Delta f_3\end{bmatrix} \tag{13}$$

It should be noted that, if the pressure is specified at one node, one equation will be dropped from the system of equations. This eliminates f_3 at that nodal point and hence the residual due to continuity of velocity is not minimized locally. To overcome this difficulty, one can assume the specified pressure values as unknown and then treat the pressure gradient across the exit and inlet boundary $[(P_{in}-P_{ex})/L]$ as a body force or make the outward force calculated from the pressure obtained from the solution be equal to the force exerted by the specified pressure values.

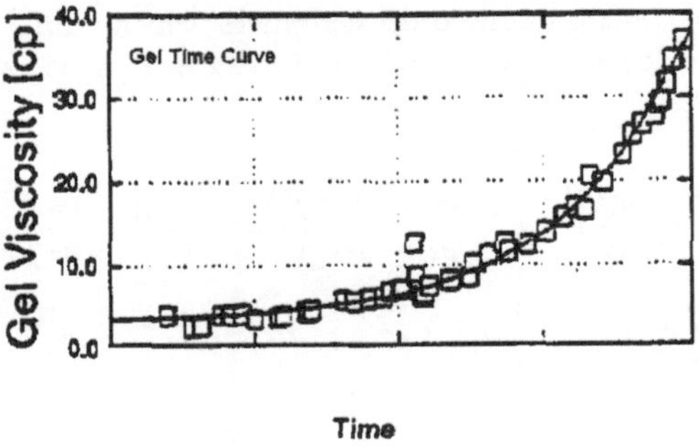

Figure 3. Variation of Gel Viscosity with Time [13]

The Solidification Process

Viscosity of CS is a function of time, silica concentration and the amount of reagents added. Since this is a theoretical investigation, only the relationship between viscosity and time is considered in this research. The variation of gel viscosity with time was obtained from experimental data and used in this research. The early study of the relationship between viscosity versus time was performed at Lawrence Berkeley National Laboratory [11]. Base on these measurements, an exponential relationship of viscosity and time was suggested by Finsterle [12]. This expression can be described as follows. During the first hour after gelling process, the initial, relatively low viscosity of the grout determines the flow behavior of the grout plume, where gravity and capillary forces affect the location and the shape of the grout plume. During the gelling process the viscosity of CS grout increases, turning the gel-water mixture into a non-Newtonian, visco-elastic fluid that eventually solidifies. The appearance of a new phase leads to change in the physical and chemical properties of contaminated soils. After the first hour, the viscosity increases rapidly and the grout starts to lose its fluid properties. The experimental result of the early gelation (during the first hour) is shown as Figure 3. The mathematical equation representing the experimental data is described as the follows.

$$\mu_{gel} = a_1 + a_2 \cdot \exp(a_3 \cdot t) \tag{14}$$

where: t =time

a_i = fitting parameters

This equation was used to design a viscous barrier for isolated of the 281-3H Retention Basin at the Savannah River Site [8] and for an injectable subsurface barrier at the Brookhaven National Laboratory site, Long Island, New York [14].

In this study the contaminated soil was idealized and modeled as a repetitive array of rigid particles extending indefinitely in all directions. It is proposed to include the variation of viscosity of pore fluid ($\varepsilon_x = \varepsilon_y = \mu$) with time rather than keeping it a constant. As the viscosity of the silica gel increases with time the above velocity field changes with many points of zero or near zero velocities close to the solid particle. The above can be viewed as coating of the soil, which results in expansion of the solid volume. With time the grout is set and a "new" porous medium is formed, which has a lower porosity, a new pore structure, and a reduced permeability. The change of viscosity with time was incorporated into the microscopic model to observe the flow pattern of CS in soil void spaces. The simulation test results are described in the following section.

Simulation Test Results

As described above, the gel is modeled as a liquid, the viscosity of which increases with time until solidification occurs. The flow pattern of colloidal silica in soil and the change of viscosity were examined. Then the movement of grout as a function of time and distance was simulated

At the beginning, colloidal silica, which consists of long chain particles in solution, is injected into soil media. During the starting period, CS grout has very low viscosity. It behaves like fluid while penetrates through soil void spaces. Navier-Strokes equation, determined the fluid flow velocity field in the unit cell. The fluid flow velocity closer to the soil surface is lower than that is away from the soil, and the velocity of fluid in the wider area of unit cell is lower than narrower area. As show in Figure 4., the velocity in the center is the highest and the velocity in the boundary of soil particle is the lowest. Thus it takes more travel time for fluid that close to soil surface than the fluid in the middle between the two soils particles.

Now considering grouting process, which contains colloidal silica particles in fluid. The gellation process begins with the initiation of grouting. As discussed above, gellation depends mainly on time. Thus CS particles that close to soil surface, which travel slower, will increase viscosity resulting in solidified CS. As discussed in the above section, the viscosity of colloidal silica is an exponential function to time (Figure 3.). At plateau point of the curve, the viscosity increases rapidly until it becomes a solid. The CS particles that require longer travel time will solidify first. This causes the increase in the solid fraction of the unit cell, decreasing the void space. The process continues until the void is filled with CS

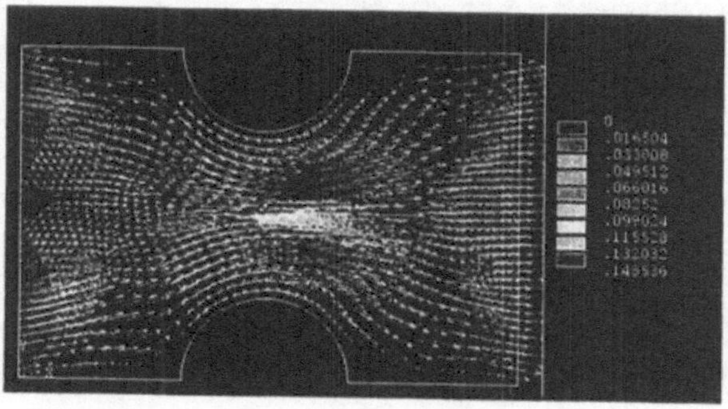

Figure 4. Velocity Profile of fluid in Unit Cell

Figure 5. Variation of Gel Viscosity with Time

gel. The increasing of CS viscosity with time is illustrated in Figure 5. Finally, the gel fills most of the voids in soil media result in partial stoppage of fluid flow through soil. This causes a drastic reduction of soil permeability.

After complete solidification, the soil exhibits new pore structure with different physical properties. The new pore structure that is predicted based on the above simulation results in the change of soils properties, such as, porosity, permeability, relative permeability and capillary pressure functions and initial liquid saturation.

SUMMARY AND CONCLUSIONS

The low-viscosity chemical grout, colloidal silica (CS), is extensively used as a grouting material in construction of grout curtains. CS with low viscosity and non-toxicity is injected into soil to stabilize fine-grained soils. It is also used as a viscous liquid barrier in some hazardous waste sites [15]. The objective of this paper is a theoretical examination of the applicability of injected CS grouts to stabilize/solidify chromium contaminated soils from Hudson County, NJ.

During construction, after liquid CS is injected into contaminated soil, it moves through the voids in the soil matrix. The viscosity is increased while CS is moving, causing a reduction of flow in the soil. The above process results in increasing in boundary layer thickness, transforming liquid CS into a gel. The gelling time and gelling mechanisms are controlled by the variation in viscosity, which depends on soil type, soil composition, concentration and chemical characteristics of CS. In this paper, the movement of CS during the grouting process as a function of time, chemistry of grout and environmental factors was modeled at the microscopic level. After injecting the grouts into soil media, the viscosity increased reducing the void spaces in soil. The soil structure changed with time due to the change of CS viscosity. This physical model described in this paper will provide design information for the real world application of CS to stabilize chromium contaminated soils.

REFERENCES

1. Noll, R. Mark., Craig Bratlett and Tina M. Dochat. 1992. "In Situ Permeability Reduction and Chemical Fixation Using Colloidal Silica." *Proceeding of the Sixth National Outdoor Action Conference on Aquifer Restoration, Ground Water Monitoring, and Geophysical Method, National Ground Water Association.* 443-457.

2. United State Environmental Protection Agency, USEPA, Office of Solid Waste and Emergency Response. March 1997. "Recent Developments for In Situ Treatment of Metal Contaminated Soils." *EPA-542-R-97-004.* Washington D.C.

3. Allan, L. M. and E. L. Kukacka, July-August 1994. "Grout-Treat Soil for Low-Permeability Barrier around Waste Landfill." *ACI material Journal.* Vol.91:355-361.

4. Watanabe et al. February 1990. "Chemical Grout for Ground Injection and Method for Accretion." *US patent 4,904,304.*

5. Persoff, Peter., John Apps, George Moridis, and Joyce M. Whang. June 1999. "Effect of Dilution and Contaminants on Sand Grout with Colloidal Silica." *Journal of Geotechnical and Geoenvironmental Engineering.* Vol.125(6):461-469.

6. Yonekura, R. and M. Miwa. May 1993. "Fundamental Properties of Sodium Silicate Based Grout." *Eleventh Southeast Asia Geotechnical Conference: May 1993.* Singapore. 439-444.

7. Persoff, Peter. et al. March 1995. "Injectable Barriers for Waste Isolation" *LBNL-36739 UC-600* Earth Science Division, Ernest Orlando Lawrence Berkeley National Laboratory.

8. Apps, A. John., Peter Persoff, Gorge J. Moridis, and Karsten Pruess. 1998. "Method for Formation of Subsurface Barriers Using Viscous Colloidal." *United State Patent 5,836,390.*

9. Moridis, G. J., A. James and C. Oldenburg. 1997. "Development of a design Package for a Viscous Barrier at the Savannah River Site." *LBNL-39487 UC-2000* Earth Science Division, Ernest Orlando Lawrence Berkeley National Laboratory.

10. Persoff, Peter., Gorge J. Moridis, John A. Apps and Karsten Pruess. 1998. "Technical Note: Evaluation Tests for Colloidal Silica for Use in Grouting Applications." *The America Society for Testing and Materials.* Vol.21(3) 264-269.

12. Finsterle, S. and C. M. Oldenburg et al. September 1996. "Mathematical Modeling of Permeation Grouting and Subsurface Barrier Performance." *LBNL-39419: UC-2000.* Earth Science Division, Ernest Orlando Lawrence Berkeley National Laboratory.

13. Moridis, G. J. et al. 1996. "A Design Study for the Isolation of the 281-3H Retention Basin at the Savannah River Site Using the Viscous Liquid Barrier Technology." *LBNL-38920 UC-2000* Earth Science Division, Ernest Orlando Lawrence Berkeley National Laboratory.

14. Morisdis J. Gorge, Stefan Finsterle and John Heiser, October 1999. "Evaluation of Alternative Designs for an Injectable Subsurface Barrier at the Brokhaven National Laboratory Site, Long Island, New York." Water Resource Research. Vol.35(10):8937-2953.

13. Moridis, G. J. et al. 1996. "A Design Study for Medium-Scale Field Demonstration of Viscous Barrier Technology." *LBNL-38916 UC-2000* Earth Science Division, Ernest Orlando Lawrence Berkeley National Laboratory.

11. Moridis, G. J. et al. 1994. "Containment of Contaminants through physical Barriers Formed from Viscous Liquids Emplaced under Controlled Viscosity Conditions," FY 1993 Annual Report *LBNL-29400,* Ernest Orlando Lawrence Berkeley National Laboratory.

15. Persoff et al. March, 1995. "Injectable Barrier for Waste Isolation." *LBNL-36739 UC-600* Earth Science Division, Ernest Orlando Lawrence Berkeley National Laboratory.

REMOVAL OF TOXIC METALS FROM CONTAMINATED WATER USING POWDER ACTIVATED ALUMINA COMBINED WITH COAGULATION/FLOCCULATION

DAVID C. STEJSKAL
Research Assistant
Auburn University
Department of Civil Engineering
238 Harbert Engineering Center
Auburn, AL 36849

TIMOTHY A. KRAMER
Assistant Professor
Texas A&M University
Civil Engineering Department
3136 TAMU
College Station, TX 77843-3136

MARK O. BARNETT
Assistant Professor
Auburn University
Department of Civil Engineering
238 Harbert Engineering Center
Auburn, AL 36849

OVERVIEW

An investigation into the removal of toxic metals from contaminated water was conducted using synthetic surface water combined with powder activated alumina (PAA) and coagulants. The mechanism used was conventional coagulation/flocculation simulated by standard jar test. Varied synthetic water contaminated with 0.5 mg/L of toxic metals was prepared for each test in six 1.0-liter beakers. The variables in this investigation were as follows: metal type (arsenic (III), lead (II), or chromium (VI)), initial pH (6.00 versus 8.00), PAA (250 mg/L versus 1000 mg/L), total suspended solids (TSS) (0 mg/L versus 100 mg/L), coagulant (aluminum sulfate versus ferric chloride), coagulant dosage (0, 10, 25, 50, 100, 150 mg/L for aluminum sulfate and 0, 5.4, 13.5, 27.0, 54.0, 81.0 mg/L for ferric chloride). Prior to testing, the pH was adjusted, and PAA, TSS, and coagulants were added to the beakers. Each test was duplicated and tested for turbidity, color, final pH, metal concentration, and residual aluminum concentration. The toxic metals were tested for removal in individual tests. The objective of this study was to determine optimum PAA dosage, coagulant, coagulant dosage and initial pH to provide the most efficient and cost effective metal removal.

The results of the investigation showed metal removal of greater than 97% (less than 15 μg/L remaining), which is below the current USEPA maximum contaminant level (MCL) for lead and chromium and within the range of the proposed arsenic MCL of 5-20 μg/L. In general, greatest removal was seen with ferric chloride as the coagulant, 1000 mg/L PAA, 150 mg/L TSS, and pH = 8.0, while removal with similar conditions, but aluminum sulfate as the coagulant produced results close to the levels achieved with ferric chloride. Furthermore, at optimum metal removal, turbidity was less than 2.0 NTU, color was less than 30 CU and residual aluminum concentration was insignificant. The results show that PAA combined with coagulation techniques is a viable and cost effective option for removal of toxic metals from contaminated water.

INTRODUCTION

One of the major factors affecting the potable water treatment industry is the presence of toxic metals in the natural source water. Toxic metals are defined as priority pollutants by the United States Environmental Protection Agency (USEPA) and must be removed from the water below stringent concentration levels prior to distribution as drinking water or released to a natural environment. As defined by the USEPA, toxic metals include: arsenic, barium, cadmium, chromium, copper, lead, mercury, and selenium. The removal of these toxins can prove to be difficult and costly with varied degrees of efficiency. Their small size and chemical composition make removal by current conventional water treatment processes unacceptable with proposed drinking water standards. Determining a process to effectively and cost efficiently remove these toxic metals is important to the safety and compliance of the water treatment industry.

Papadiamas *et al.* (1997) [1] identified six treatment methods that have been proven effective in toxic metals removal from contaminated water. These processes are:

- Coagulation and filtration with or without sedimentation
- Coagulation assisted micro-filtration membrane separation
- Lime softening
- Sorption on granular activated alumina
- Sorption on ion exchange resins
- Reverse osmosis membrane separation

However, each of these traditional methods of treatment involve substantial cost and are often difficult to properly control, especially for small treatment systems and hazardous clean-up operations. It is noteworthy in the study of Papadiamas *et al* [1]. that alumina sorption was found to be the most effective toxic metals removal technology.

Activated alumina (solid Al_2O_3) has been shown to be an effective sorbent for toxic metals (Clifford (1990)[2], Gupta and Chen (1978)[3], and Fox (1989)[4]). Alumina is the stable form of aluminum found in the environment and is one of the most common minerals found in nature. Prior research in using alumina as a removal technology for the treatment of water has focused on passing the treated water through fixed beds of large sized granules of alumina. Although alumina is inexpensive (less than $0.30 per lb. for pure Al_2O_3) the cost associated with regenerating and handling the granular beds of alumina has limited its use. Decreasing the size of the alumina particles increases both the sorption rate and capacity of metal removal. The decrease in particle size increases the surface area per gram of alumina added thus increasing the area available for sorption to occur (Wootton *et al.* (1998)[5]). However the subsequent removal of the fine particle alumina from treated water has not been investigated thoroughly.

In the research that follows, the removal of toxic metals arsenic, lead, and chromium will be investigated. The selection of these metals was based on their toxicity and concern to regulatory agencies and the varying oxidation state/speciation and removal efficiency using activated alumina. The use of powder activated alumina combined the coagulation/flocculation process was studied to determine the viability of its use to remove toxic metals while producing a treated water that meets water regulations.

Previous research by Wootton et al (1998)[5] showed that metals adsorb to the surface of powdered activated alumina. One of the major drawbacks associated with the addition of PAA was the need to remove the fine particle alumina from the water. Ultrafiltration has shown to be an effective way to remove the alumina from an aqueous solution, but the ability to use PAA in conventional water treatment would be a major breakthrough. Investigations done by the researchers prior to this work showed that the addition of the finely divided alumina aided coagulation. The alumina, when placed in an aqueous environment, develops a charged surface given the PAA colloidal properties. The charged surface will attract colloids of opposite charge, thus acting as flocculating nuclei. Flocculation that is initiated by the addition of small particles is termed ballasted flocculation and the addition of alumina powder will act as such ballast.

The combination of powdered activated alumina with conventional water treatment processes, such as coagulation/flocculation, has not been previously studied. In conventional water treatment, coagulant chemicals, typically alum ($Al_2(SO_4)_3$) or ferric chloride ($FeCl_3$) are added to the treated source water, rapidly mixed, and then slowly agitated for a period of time to induce flocculation of the turbidity causing compounds. The resulting floc is removed by sedimentation followed by filtration. The addition of powdered activated carbon (PAC) in the rapid mix phase of conventional treatment for the removal of soluble organics is a common practice (Adham (1991)[6], Adham (1993)[7], and Najm (1990)[8]). The use of PAC as a simple retrofit to an established treatment plant is easy to apply and has proven to be cost effective. Furthermore, the research of Yallaly (1999)[9] showed that arsenic rapidly adsorbs to PAA and the subsequent removal of arsenic rich PAA successfully removed the arsenic levels to acceptable limits. A rational extension is to assume that a similar technique using powdered activated alumina and conventional water treatment would be a cost effective technology for the removal of harmful toxic metals.

EXPERIMENTAL METHODS

Sample Preparation

The PAA was prepared by crushing Fisher Scientific γ–phase granular activated alumina using a U.S. Millstone pulverizer. The alumina was then separated using an ASTM standard sieve set and sieve shaker. All the alumina retained on the #200 sieve (75 μm) and passing the #150 sieve (150 μm) was used in the experiments. The retained alumina was oven dried at 105 °C for 24 hours prior to being weighed so that the sample weight was due solely to the alumina and not from water absorbed from the atmosphere. The total suspended solids were derived from native Alabama soil, which has high iron content. The solids were oven dried at 105 °C for 24 hours and then crushed and sieved. The solids passing the #200 sieve (75 μm) were collected and used for the suspended solids addition. Arsenic, lead, and chromium stock solution were prepared following the *Standard Methods for the Examination of Water and Wastewater* (Method 3114 for As(III) and Method 3111 for Cr(VI) and Pb(II))(1998)[10]. The stock solutions were then used to spike the experiment beakers with 500 μg/L. The alum solution was prepared using granular aluminum sulfate ($Al_2(SO_4)_3 + 14.3\ H_2O$), and the ferric chloride solution was prepared from anhydrous $FeCl_3$.

Test Matrix

For each metal tested, the following matrix was used. The 2^4 Factorial Design was adapted from *Statistics for Experiments* (Box *et al.*, 1978)[11].

TABLE I. EXPERIMENTAL DESIGN MATRIX

Experiment Number	Alumina Addition	Coagulant	Total Suspended Solids	pH
1	-	+	+	-
2	-	+	-	+
3	-	-	+	+
4	-	-	-	-
5	+	+	+	+
6	+	+	-	-
7	+	-	+	-
8	+	-	-	+

Variables:

Alumina Addition	+ = 1000 mg/L	- = 250 mg/L
Coagulant	+ = Alum	- = Ferric Chloride
Total Suspended Solids	+ = 150 mg/L	- = 0 mg/L

pH	+ = 8.00	- = 6.00

Test Preparations

A series of test were run on a traditional jar test apparatus. Each sample was prepared in a 1-liter beaker spiked with 500 µg/L of the metal (As(III) as As_2O_3, Pb(II) as $Pb(NO_3)_2$, and Cr(VI) as CrO_3) and adjusted to an initial pH using 0.1 N sodium hydroxide and 0.1 N nitric acid solutions. Once the pH was adjusted, the beakers were placed on the jar test apparatus and stirred at 100 rpm. The coagulant dosages were taken from concentrated aluminum sulfate and ferric chloride solutions and measured to the increasing dosages for each jar. The PAA and TSS were weighed on an analytical top load balance. Once all additions were prepared they were added simultaneously. Each test run was preformed with 30 seconds of rapid mix (100 rpm), 15 minutes of slow stirring (25 rpm), followed by 30 minutes of settling. A sample was taken from each beaker and analyzed for turbidity, pH, color, metal concentration, and residual aluminum.

Sample Analysis

The samples were collected in two 15 ml tubes and analyzed for pH using Orion Model EA 940 pH meter and turbidity and color using a HACH 2100AN turbidimeter immediately. The metal concentrations were analyzed within a week using Perkin-Elmer Model 3100 atomic sorption spectrophotometer equipped with a HGA 600 graphite furnace. One 15 ml tube was filtered with a 0.45 µm membrane filter and analyzed on an inductively coupled plasma analyzer (ICAP) for residual aluminum.

RESULTS AND DISCUSSIONS

Metals Removal

All three metals, arsenic(III), lead(II), and chromium(VI), showed significant removal with the addition of PAA and coagulants. Total removal ranged from 75% to over 98% at optimum conditions for each test run. The best results were achieved when using ferric chloride as a coagulant, with pH = 8.0, 1000 mg/L PAA, and 150 mg/L of TSS. These results agree with expectations except for better metal removal with an initial pH of 8.0 versus 6.0. By studying the final pH data, it can be seen that with the addition of coagulants, pH decrease as expected. This results in the pH for the test with initial pH = 8.0 falling between 6.0-7.0 when the optimum metal removal efficiency was seen. With an initial pH = 6.0, the addition of small amounts of coagulant causes the pH to be lower than acceptable ranges for coagulation (5.8-8.0 for alum and 4.5-8.5 for ferric chloride) (Amirtharajah and O'Melia, 1990)[12], hindering coagulation and thus the removal of metals.

When studying the effect of the PAA on removal, in all cases, the higher activated alumina concentration showed better removal results, which was expected. With the increase of activated alumina, more sites were available for adsorption thus more metals could be adsorbed. The toxic metals attach to the surface of the alumina, and once the alumina settles out or is remove by coagulation, the toxic metals are removed as well (Figure 1 and Figure 2).

In the research of Jeffcoat (2000)[13], it was shown total suspended solids positively impacted the removal of metals from water. The total suspended solids used in this research were derived from natural Alabama clay soil, which has high iron content. The iron content in the suspended solids aids in the removal of metals by forming additional sites for metal sorption. The 150 mg/L suspended solids added was an average TSS reading for local rivers and streams in the Lower Chattahoochee River valley. The test results seem to support the work of Jeffcoat [13]. Tests run with only suspended solids showed some removal of toxic metals, but with low efficiency. The removal of suspended solids does play a part in metal removal, but sorption onto the surface yields only minor improvements in toxic metals removal efficiencies.

Generally, ferric salts are known as a better removal of metals than alum. While some researchers disagree on why ferric salt removal is better, this research seems to support the claim with the observed results showing slightly better metal removal for test in which ferric chloride is the coagulant (Figure 1). Optimum metal removal with alum was seen at near 95% removal (Figure 3) while ferric chloride reached levels over 98% removal. It is the opinion of the researchers that ferric chloride is a better overall removal of metals but alum's significantly better results in turbidity removal compensate for its lower metal removal efficiency in this research. With the addition of PAA to the jars, the adsorption of metals to the alumina surface is the main catalyst for removal. The subsequent removal of the metal rich alumina greatly impacts total metal removal. Aluminum sulfate coagulation provides a much better final turbidity reading than ferric chloride. It is reasonable to assume that the lower turbidity readings are proportional to better dissolved and soluble alumina removal and thus lower metal removal. Yet, even with this extra removal aid, ferric chloride still consistently produces better toxic metals removal.

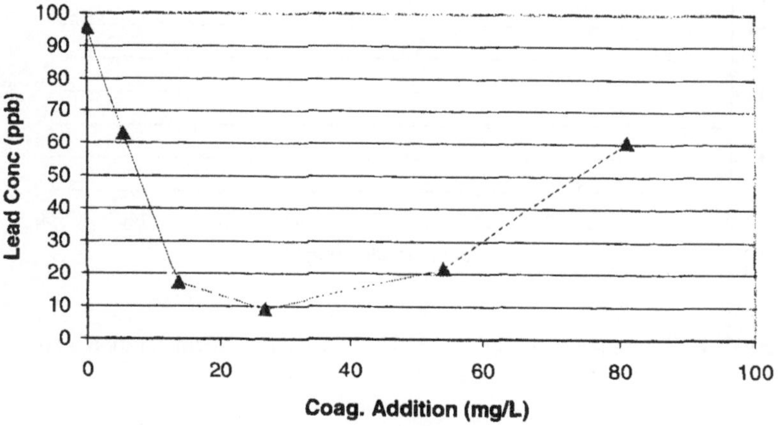

Figure 1: Pb(II) with 1000 mg/L Alumina, 150 mg/L TSS, ferric chloride addition, and pH=8 with an initial 500 ppb Pb(II) concentration

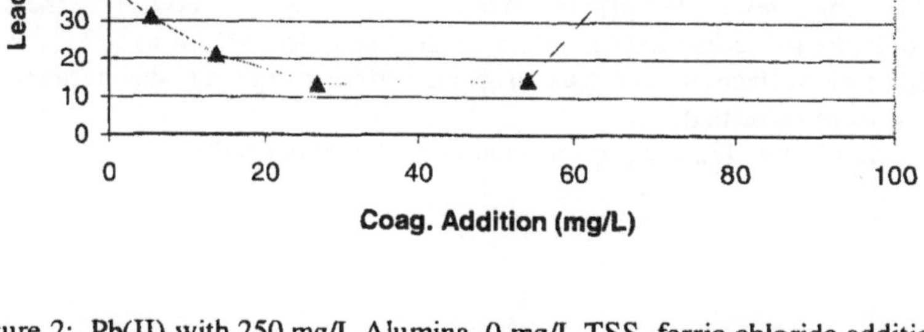

Figure 2: Pb(II) with 250 mg/L Alumina, 0 mg/L TSS, ferric chloride addition, and pH=8 with an initial 500 ppb Pb(II) concentration

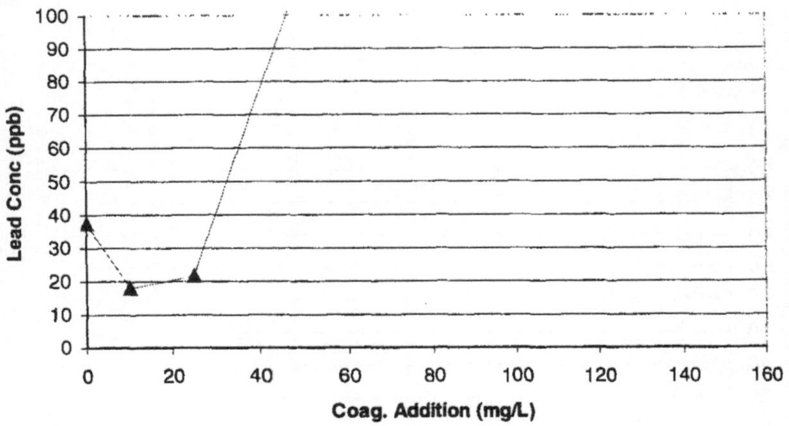

Figure 3: Pb(II) with 1000 mg/L Alumina, 0 mg/L TSS, alum addition, and pH=8 with 500 ppb initial Pb(II) concentration

Turbidity and Color

As stated earlier, turbidity and color were two major variables studied in this research. Providing water that not only remove metals, but also met water quality standards for turbidity was important. Turbidity readings ranged from 0.34 – 37.8 NTU for each tests optimum results. As expected, the use of alum as a coagulant was significantly better at removing turbidity than ferric chloride. The best results with alum as the coagulant were 0.34 NTU (Figure 4) while ferric chloride best results were around 1.0 NTU (Figure 5). Both optimum conditions occurred with 250 mg/L TSS and initial pH = 8.0. These better results can be attributed to alum addition forming a greater variety of aluminum hydroxide complexes while ferric chloride is mostly present as $Fe(OH)_3^{o}$. The variety of hydroxide complexes allows for a greater variety of ions to be attracted to the surface, reducing electrostatic surface charge, and then forming floc particles to be settled.

Color results were directly proportional to the turbidity results.

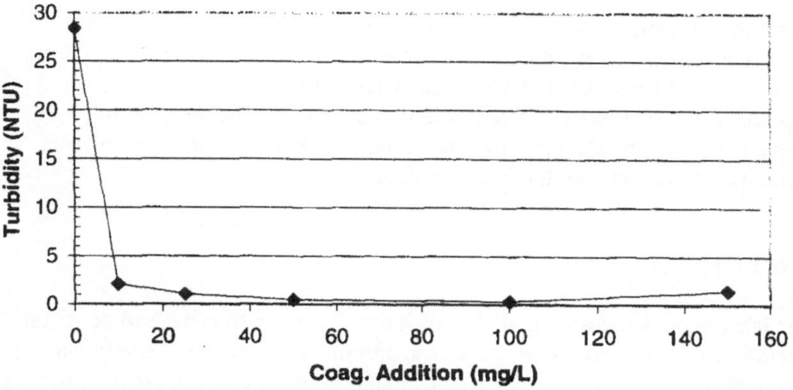

Figure 4: As(III) with 250 mg/L Alumina, 150 mg/L TSS, alum addition, and pH = 8

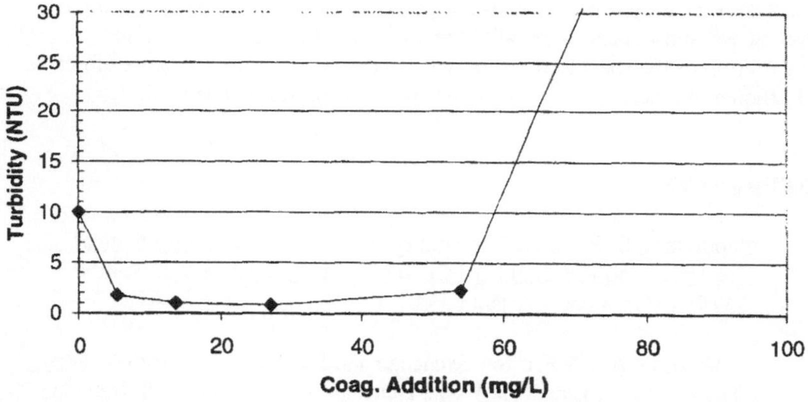

Figure 5: As(III) with 250 mg/L Alumina, 0 mg/L TSS, ferric chloride addition, and pH=8

Residual Aluminum

With the recent concerns of a possible link between aluminum consumption and Alzheimer's Disease, the residual aluminum concentration is an important variable to consider. Also, since metal removal is dependant on increasing activated alumina addition, aluminum present in the filtered water was studied. In all cases, the filtered samples tested measured below 1 ppm, which is acceptable for potable water. The low levels suggest that the removal of dissolved and suspended aluminum is achieved by the coagulation/flocculation system in combination with the filtration step.

CONCLUSIONS

The use of powdered activated alumina has been shown to reduce toxic metal concentration in synthetic water yet its use presents a balance between finished water quality and metal removal. As the amount of PAA added increases, the removal of the toxic metals increases as well. However, increasing the amount of PAA increases the water turbidity and color, causing the water quality to decrease and increases the need for coagulant addition. Furthermore, high levels of PAA cause more solid sedimentation to occur, which must eventually be removed and disposed. Therefore, a delicate balance is present which must be tailored to the individual need of each system, plant, or processes.

This research supports the use of PAA with conventional water treatment processes for the removal of toxic metals. However further research needs to be undertaken to study some areas of this process. Research needs to examine the removal and subsequent disposal of the metal rich spent alumina. Also regeneration of the spent alumina powder to determine if this is a cost effective and efficient method to reuse to alumina needs to be studied as well.

REFERENCES:

1. Papdiamas, S. P., Z. K. Chowdhury, et al. (1997). Development of Treatment and Residual Cost Curves for Arsenic Removal. Annual AWWA Conference, Atlanta, GA.

2. Clifford, D. A. (1990). Ion Exchange and Inorganic Adsorption. Water Quality and Treatment. AWWA. Denver, CO, McGraw-Hill, Inc: 561-640.

3. Gupta, S. K. and K. Y. Chen (1978). "Arsenic Removal by Adsorption." Journal WPFC: 493.

4. Fox, K. R. (1989). "Field Experience with Point-of-Use Treatment Systems for Arsenic Removal." Journal AWWA **81**(2): 94.

5. Wootton, P., D. M. Cheek, et al. (1998). Treatment of Heavy Metals Contaminated Waste Water Using Powdered and Granular Alumina.

31th Annual Mid-Atlantic Industrial and Hazardous Waste Conference, Villinova University, Philadelphia, PA.

6. Adham, S. S., V. L. Snoeyink, et al. (1991). "Predicting and Verifying Organics Removal by PAC in a Ultra-filtration System." Journal AWWA **83**(12): 81.

7. Adham, S. S., V. L. Snoeyink, et al. (1993). "Predicting and Verifying TOC Removal by PAC in Pilot-Scale UF System." Journal AWWA **85**(12): 58.

8. Najm, I. N., V. L. Snoeyink, et al. (1990). "Effect of Particle Size and Background Natural Organics on the Adsorption Efficiency of PAC." Journal AWWA **82**(1): 65.

9. Yallaly, B. (1999). Arsenic Removal from Surface Water Using Ultra-filtration and Powdered Activated Alumina. Masters Thesis Department of Civil Engineering. Auburn, AL, Auburn University.

10. Clesceri, L. S., A. E. Greenberg, et al. (1998). Standard Methods for the Examination of Water and Wastewater. Washington, D. C., American Public Health Association.

11. Box, G. E. P., W. G. Hunter, et al. (1978). Statistics for Experiments. New York, NY, John Wiley and Sons, Inc.

12. Amirtharajah, A. and C. R. O'Melia (1990). Coagulation Processes: Destabilization, Mixing, and Flocculation. Water Quality and Treatment. AWWA. Denver, CO, McGraw-Hill, Inc.: 269-365.

13. Jeffcoat, S. (2000). Removal of Toxic Metals and Reduced Membrane Fouling Using Powdered Activated Alumina. Masters Thesis Department of Civil Engineering. Auburn, AL, Auburn University.

THE SO_2/O_2 SYSTEM AS A NOVEL APPROACH FOR CLEANING CONTAMINATED SEDIMENTS

MAURIZIO MAREZIO-BERTINI
Earth and Environmental Engineering Department
Henry Krumb School of Mines
Columbia University in the City of New York
New York, NY 10027

PAUL F. DUBY
Earth and Environmental Engineering Department
Henry Krumb School of Mines
Columbia University in the City of New York
New York, NY 10027

ABSTRACT

Contaminated sediments become an issue when environmental or human risk is identified and navigational needs require that contaminated sediments be dredged from shipping channels. In the United States, 14 to 28 million cubic yards of sediments must be managed every year and decontamination is the necessary step before disposal or reuse as a secondary material.

Since the SO_2/O_2 system has been successfully applied to the destruction of cyanide in a safe and reliable manner, it is worthwhile to investigate its possible use for the remediation of contaminated sediments. It was found that the metal content of contaminated sediments decreased significantly after treatment in a NaOH solution sparged with the SO_2/O_2 mixture. In particular, metals like Cu, Cd, Pb and Cr were removed by this procedure. The efficiency of the process was function of the duration of the treatment, its temperature and the final pH. By increasing the temperature, the metal removal efficiency could be improved, with the exception of Cu. A final acidic pH produced higher metal recoveries than a basic pH. Under optimum conditions, a process run at 80 °C, for 120 minutes with a NaOH concentration of 0.5 M was able to remove 90% of Pb and Cr, and 70% of Cu. A treatment for 300 minutes, at 25 °C with a final pH equal to 1.5 completely removed Cd. Sedimentation tests indicated that the particle size distribution of the sediments changed after treatment with the appearance of a finer size; it can be postulated that organic compounds, binding these particles together, were destroyed during the treatment.

INTRODUCTION

The U.S. Environmental Protection Agency (EPA) defines sediments as soil, sand, and minerals washed from land into water usually after rain. In a broader definition, the term "sediment" encompasses the various materials that settle to the bottom of any water body [1].

Contaminated sediments become an issue when environmental or human risk is identified and when navigational needs require that contaminated sediments be dredged from shipping channels. In 1997, it was estimated that 14 to 28 million cubic yards of contaminated sediments must be managed every year in the U.S.

Contamination of sediments results from human activities and are usually classified from their origin as:

- point sources such as municipal and industrial effluents,
- non-point sources such as agricultural runoff, soil entrainment, or airborne particles,
- other sources such as spills, contaminated ground water infiltration, and dumping.

Sediments that have to be dredged from estuaries to keep navigation lanes open are strongly contaminated by human activities. They may no longer be dumped in the ocean without prior decontamination. They must be shipped to a disposal site or treated before dumping. With further treatment, they may be converted for use, for instance, as construction materials. An example is the problem now facing the Port of New York and New Jersey in managing the amount of dredge material needed to keep navigation channels open. Based on current testing criteria, only about 25% of the material is suitable for ocean dumping at an estimated cost of $5-$10 per cu yd. About 9% qualifies for "restricted" ocean disposal requiring special capping at a cost of $35 per cu yd. About 66% fails ocean disposal criteria and requires treatment or shipping to a disposal facility. In one case, approximately $118 per cu yd was paid to ship material to a facility in Utah [2].

Present available technologies are either slow or fairly expensive. Typical treatment schemes include removal, transport, dewatering, and size classification to remove oversized material. They are then followed by preconditioning, principal treatment, and solid residue treatment, each of them producing a side stream to be treated before disposal. The successfully tested principal treatments can be classified as:

- biological treatment,
- dechlorination,
- solvent extraction,
- soil washing,
- thermal desorption,
- incineration [1].

These typical treatment schemes can be modified by adding before dewatering, or in combination with it, a leaching (or washing) step to remove some of the oil and grease adsorbed on the fine particles surface, leach contaminating metals (e.g. Cu, Cr, Cd and Pb), and oxidize some refractory organic compounds.

THE SO_2/O_2 SYSTEM

The oxidizing capability of the SO_2/O_2 system is still a surprising one, being that SO_2 is a well-known reducing agent and O_2 a mild oxidizer. The fact that small amounts of sulfur dioxide significantly increase the rate of oxidation of ferrous iron by oxygen has been known for some time [3,4]. The overall stochiometry of the reaction is:

$$2Fe^{++} + SO_2 + O_2 = 2Fe^{+++} + SO_4^{--} \qquad (1)$$

The oxidation of sulfite ions by oxygen has been reported to be a chain reaction [5]. It is catalyzed in the presence of transition metals. Although data are scarce in the literature, the existence of sulfite-metal complexes is recognized and it has been suggested that they enhance the formation of unstable intermediate. The more stable monopersulfiric ion, SO_5^{2-}or Caro's acid, has been postulated [5] The SO_2/O_2 system has found applications for oxidations such as Fe^{2+} to Fe^{3+}, Ni^{2+} to Ni^{3+}, and cyanide to cyanate [6,7].

The purpose of this paper is to introduce a new possible application for the SO_2/O_2 system in the treatment of contaminated sediments, as a leaching step in which a small amount of sulfur dioxide gas will be mixed with oxygen and fed to a slurry of contaminated sediments to destroy organic and metallic contaminants.

EXPERIMENTAL

Tests were carried out in a five-neck 1000 ml glass kettle, at temperatures between 25 and 80 °C. The total gas flow rate was 1200 ml/min and had a composition of 98% O_2 and 2% SO_2. A slurry of 2% in solid was obtained by mixing sediments in 500 ml of a NaOH solution and stirred at 500 rpm in order to obtain the complete suspension of the particles and optimal mixing conditions. The slurry's pH and redox potential (vs. calomel electrode) were monitored during the whole duration of the experiment. The duration of each experiment varied between 90 and 300 minutes.

The effect of pH on metals removal was investigated by varying the concentration of the NaOH solution between 10^{-4} and 0.5 M. Usually pH was left drifting towards acidic levels but, for 2 tests, it was controlled with addition of concentrated NH_4OH or NaOH.

Metal content in solution was determined by atomic adsorption spectrometry. Metal levels in sediments and residues were obtained according to the ASTM D3974-81 procedure. Average values for Cu, Cd, Pb and Cr in untreated sediments were found to be, respectively, 0.019%, 0.0009%, 0.03% and 0.0131

%. Material balances carried out on Cu, Cd and Pb produced, respectively, an average value of 99%, 100% and 101%.

RESULTS AND DISCUSSION

Effect of initial NaOH concentration

The effect of initial NaOH concentration on metals recovery efficiency was investigated by running tests at different concentration at 25 °C. The results are summarized in Table 1 and plotted in Figure 1 and Figure 2.

Metal recovery is only slightly influenced by the NaOH concentration. In fact, when NaOH concentration rises from 10^{-4} M to 10^{-3} M, Cu recovery is negatively influenced, Cd recovery is unaffected and Pb recovery is improved. When pH is controlled with addition of concentrated NH_4OH, all the metal recoveries are negatively influenced. For all the tests in Figure 1 a material balance was carried out yielding close to 100%.

TABLE 1. EXPERIMENTAL CONDITIONS AND METAL RECOVERIES.

Time (min)	NaOH (M)	Initial pH	Initial pH	Recovery (%)			
				Cu	Cd	Pb	Cr
90	0.5	12.0	2.0	54.4	78.9	30.2	NA
120	0.5	13.0	2.0	33.7	NA	40.8	42.6
120	0.5	10.0	1.4	37.0	NA	42.0	51.0
120	0.5	10.0	9.95*	40.0	NA	38.0	18.0
300	0.001	10.0	1.54	34.0	99.7	74.8	48.9
300	0.0001	8.3	1.27	46.4	100.0	56.2	56.7
300	0.0001	10.0	9.1**	44.3	59.3	48.2	56.2

* pH controlled with NaOH

** pH controlled with concentrated NH_4OH

Effect of final pH

Tests were run at higher NaOH concentration and shorter time to further investigate the effect of final pH and duration. These tests confirmed, as shown in Figure 2, that high recoveries were achieved when pH was left drifting towards acidic levels. It should be noticed that in only 120 minutes, the recoveries for Cu, Cd and Pb were more than half the recoveries obtained after 300 minutes and lower NaOH concentrations. From the results obtained, it can be concluded that in order to achieve an efficient metal removal, pH has to decrease to acidic values and treatment should last at least 120 minute.

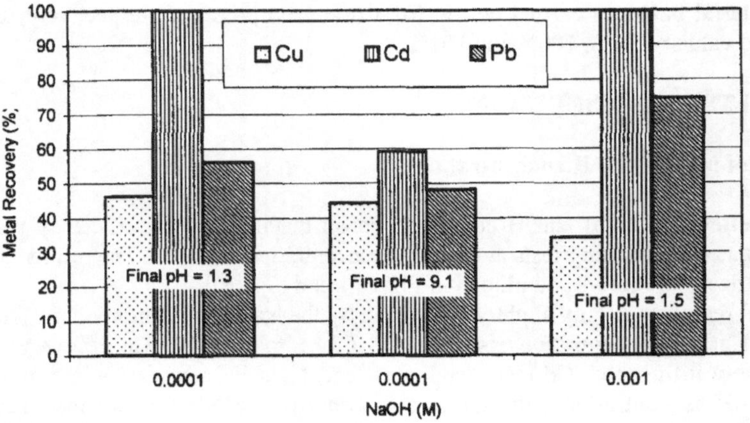

Figure 1. Metal recovery as a function of initial NaOH, at 25 °C, after 300 minutes of treatment.

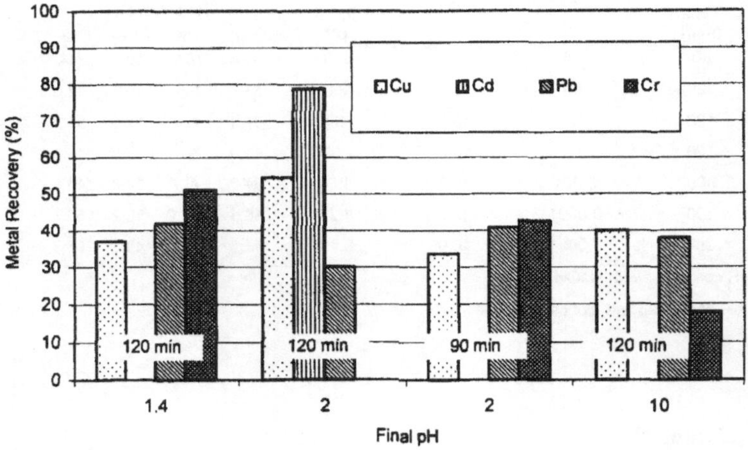

Figure 2. Metal recovery as a function of final pH, for an initial 0.5M NaOH concentration

Effect of temperature

The effect of temperature on metal removal was also investigated, since it is well known that the SO_2/O_2 system performs better at higher temperature. A set of experiments was run at 25, 55 and 80 °C, with a NaOH concentration of 0.5 M for 120 minutes. During these experiments, pH was not controlled and was left drifting towards acidic values. Since Cd was previously found to be completely removed at 25 °C, it was not investigated. The results for Cu, Pb and Cr are shown in Figure 3. An increase in temperature produces an increase in metals removal.

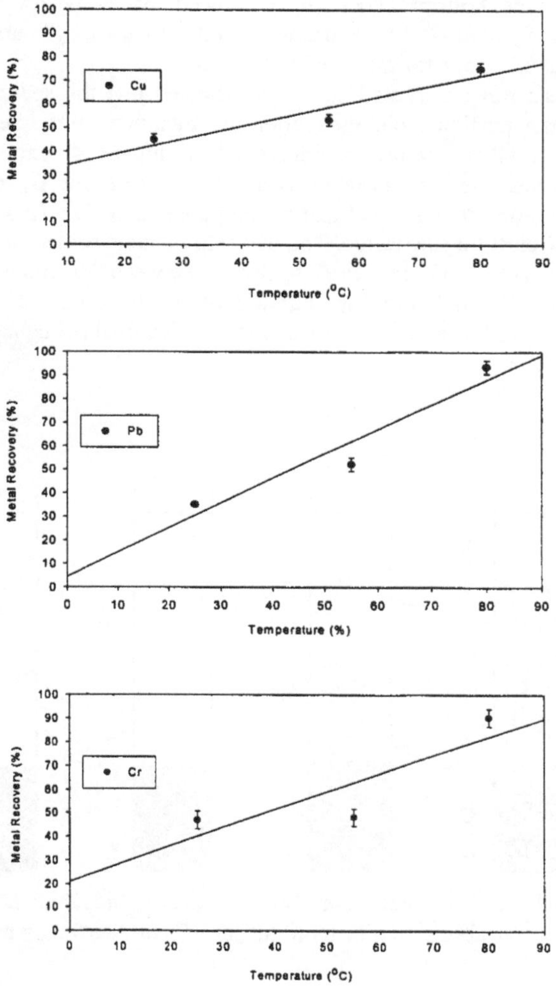

Figure 3. Recovery of metals as a function of temperature.

Sedimentation tests

Sedimentation tests were run with untreated and treated sediments to determine the settling characteristics of the particles and the changes occurring after treatment. Samples were introduced into a sedimentation jar (18.42 cm in height, 2 cm internal diameter) filled with distilled water to the reference mark. Before starting the sedimentation test, the jar was shaken for 5 minutes to ensure a satisfactory suspension of the sediments. The untreated sediments settled by forming a continuous settling column with a clear supernatant above it and a compact deposit at the bottom. These tests showed that a narrow range of particle sizes exists. Typical settling times were in the order of 24 hours for wet sediments and about one hour for air-dried sediments.

Sedimentation tests run on treated sediments showed that the sedimentation characteristics of the particles changed after the treatment with the SO_2/O_2 system. The sedimentation behavior, illustrated in Figure 4, exhibits four different settling zones: clear supernatant (zone A), slurry (zone B), region of variable size distribution (zone C) and fast-settling zone (zone D). This behavior is typical of a wider particle size range. The fast settling zone settled in about 14 minutes. The slurry zone (silt particles) settled in between 30 minutes and 8 hours, depending on the pH and the duration of the treatment. The fastest settling rate was observed after a 5 hours treatment with a final pH equal to 9.95.

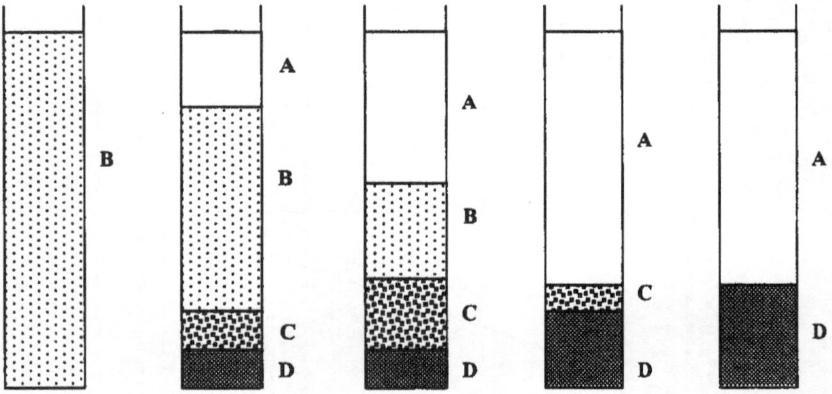

Figure 4. Batch sedimentation behavior. Zone A: clear liquid; Zone B: slurry; Zone C: region of variable size distribution; Zone D: fast-settling particles.

CONCLUSION

In conclusion the following can be stated:

- The SO_2/O_2 system is able to remove metals contaminated sediments;
- The efficiency of the process depends more on the final pH than on the initial NaOH concentration;
- Removal efficiency for Pb and Cr improves by raising the temperature but Cu is not affected;
- A treatment for 300 minutes, at 25 °C with a final pH equal to 1.5 completely removes Cd.
- A treatment for 120 minutes, at 80 °C with a NaOH concentration of 0.5 M removes 90% of Pb and Cr and 70% of Cu.

REFERNCES

1. EPA. 1992. *Selecting Remediation Techniques For Contaminated Sediments.* USEPA 823-B93-001.

2. U.S. Army Corps of Engineers. 1996. *Dredged Material Management Plan for the Port of New York and New Jersey.* Interim Report.

3. Leaver, E.S., R.V. Thurston. 1923. *Ferric Sulfate and Sulfuric Acid from Sulfur Dioxide and Air.* U.S. Bureau of Mines, R.I. 2556.

4. Ralston, D.C., C.G. Maier. 1927. *The Ferric Sulfate - Sulfuric Acid Process.* U.S. Bureau of Mines, Bull. 260.

5. Botteger, W.W., E. Thoma. 1936. "Clarification of the Wicke Reaction," *J. Prakt. Chem.*, 14T: 11-21.

6. Devuyst, E.A., A. Mosoiu, E. Krause. 1982. "Oxidizing Properties and Applications of the SO_2/O_2 System". Hydrometallurgy: Research, Development and Plant Practice. The Metallurgical Society of AIME: 291-403.

7. Arias, J., P. Duby. 1993. "Kinetics of Pyrite Oxidation in Acidified Ferrous Sulfate Solutions in the Presence of Oxygen-Sulfur Dioxide Mixtures". SME Preprint, No. 93-48.

SYMPOSIUM 3:

Bioremediation

IN SITU

ANAEROBIC DECHLORINATION OF TCE IN GROUNDWATER

VIA BIOAUGMENTATION

David E. Ellis, DuPont Company
Edward J. Lutz, DuPont Company
Ronald J. Buchanan, Jr., DuPont Company*
Craig L. Bartlett, DuPont Company
Edwin Hendrickson, DuPont Company

ABSTRACT

An anaerobic bioaugmentation field pilot system was implemented on a trichloroethene (TCE)-contaminated aquifer at Dover Air Force Base (AFB), Delaware, using a microbial enrichment culture capable of dechlorinating TCE to ethene. An insitu, hydraulically controlled pilot system 20 ft by 60 ft was established 48 ft below ground surface in an alluvial aquifer to introduce nutrients and electron donor into the groundwater. TCE and cis 1,2-dichloroethene (cDCE) concentrations in groundwater averaged 4,800 and 1,200 micrograms per liter (µg/L) prior to treatment. The insitu pilot system operated for a total of 568 days. Field results after nine months of pilot operation confirmed previous laboratory work that indicated dechlorination would not proceed past cDCE.

A non-native ethene-forming microbial enrichment culture from the Department of Energy?s (DOE?s) Pinellas Florida site was obtained, cultured and injected into the pilot area under anaerobic conditions. Subsequent to a lag period of approximately three months, vinyl chloride and ethene appeared in pilot system monitoring wells. The injected dechlorinating culture survived and was transported throughout the pilot area. After sixteen months of operation, TCE and cDCE were converted to the totally non-chlorinated end product, ethene.

KEYWORDS: Bioremediation, , trichloroethene, bioaugmentation, anaerobic

* Contact: Ronald J. Buchanan,PhD, Principal C

MODELING CHROMIUM (VI) REDUCTION AND PHENOL DEGRADATION IN A PACKED-BED BIOREACTOR: TRANSIENT STATE MODEL

YI-TIN WANG
Civil Engineering Department
University of Kentucky
161 Oliver Raymond Building
Lexington, KY 40506

EVANS M. NKHALAMBAYAUSI-CHIRWA
Howard University
Department of Civil Engineering
2300 Sixth Street NW
Washington, DC 20059

INTRODUCTION

Hexavalent chromium, Cr(VI), is found together with a variety of aromatic compounds in Cr(VI) contaminated environments [*1*]. Cr(VI) and most aromatic compounds are carcinogenic to mammals including humans and mutagenic to microorganisms. Simultaneous Cr(VI) reduction and phenol degradation has been studied in batch, suspended growth CSTR, and biofilm reactor systems using a coculture of *Pseudomonas putida* DMP-1 and *Escherichia coli* ATCC 33456 [*2,3*]. A batch kinetic model was earlier developed to simulate simultaneous Cr(VI) reduction and phenol degradation based on batch coculture studies [*4*]. However, the model presented in this paper is the first to predict simultaneous Cr(VI) reduction and phenol degradation in a coculture biofilm system. The model utilizes fully adaptive boundary conditions at the liquid/biofilm and biofilm/substratum (glass bead) interfaces and a diffusive biofilm concept with respect to particulate matter to account for changes in biofilm density within the biofilm.

MATERIALS AND METHODS

Coculture and Bioreactor Setup

E. coli ATCC 33456 was originally purchased from the American Type Culture Collection. *Pseudomonas putida* DMP-1 was a previous isolate in our laboratory and was identified as *P. putida* B using the Biolog Bacteria Identification System [*5*]. The reactor was constructed and operated as described by Nkhalambayausi-Chirwa and Wang, 2001 [*3*]. The reactor consisted of a 20 cm × 3.8 cm i.d. Pyrex glass (Corning Glassware Co.) column packed with a total of 11101, 3 mm diameter glass beads (Figure 1). Influent feed hydraulic retention times (HRTs) of 24, 11.4, 6, and 18 h were used together with a high recycle rate (300 × Q), where Q is the influent feed rate (L^3T^{-1}).

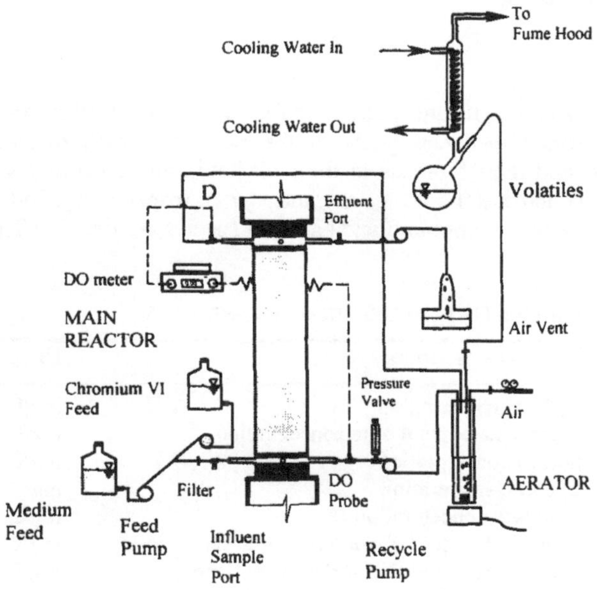

Figure 1. Schematic of fixed-film coculture bioreactor setup.

Analytical Methods

Cr(VI) and Phenol: Cr(VI) was measured by a Milton Roy spectrophotometer (Spectronic 1201, Milton Roy, NY) at a wavelength of 540 nm in H_2SO_4 acidified samples reacted with *S*-diphenyl-carbazide [*6*]. Phenol was measured by methylene-chloride extraction method [*3*] followed by injection in GC (Varian Model 3400, Palo Alto, CA) equipped with a flame ionization detector (FID) and a 15 m by 0.53 mm ID DB-5 capillary column (J & W Scientific Co., Folsom, CA) operating under detector and column temperatures of 200°C.

Metabolites: Metabolites formed from phenol degradation were determined by GC/MS (Finningan MAT INCOS 50, San Jose, CA) and by scanning spectrophotometry (Sprectronic 1201, Milton Roy, Rochester, NY) [*3*,*7*].

Biomass: Viable attached biomass was determined by serial dilution of cells washed from glass beads from three different locations at the top and bottom of the reactor (4 beads per location) followed by a pour-plate count procedure on MacConkey Agar [*8*,*9*].

MODEL DEVELOPMENT

Model Variables

The dissolved species concentrations in bulk liquid and biofilm zones are represented by the vectors $u_B(t)$ and $u_f(t,z)$, for the bulk liquid and biofilm zones. The viable biomass and inert biomass in the bulk liquid and biofilm zones are represented by $\mathbf{x}_{aB}(t)$ and $\mathbf{x}_{iB}(t)$, and $\mathbf{x}_{af}(t,z)$ and $\mathbf{x}_{if}(t,z)$, respectively. Individual variables representing each compound and bacterial species are listed in Table I.

TABLE I. LIST OF DEPENDENT VARIABLES

Symbol*	Variable Description	Units
$\mathbf{u}_c$	Cr(VI) concentration	mg/L
$\mathbf{u}_u$	intermediate metabolite concentration	mg/L
$\mathbf{u}_p$	phenol concentration	mg/L
$\mathbf{x}_e$	*E. coli* concentration	mg/L
$\mathbf{x}_p$	*P. putida* concentration	mg/L
$\mathbf{x}_{eB}$	viable bulk liquid *E. coli* concentration	mg/L
$\mathbf{x}_{pB}$	viable bulk liquid *P. putida* concentration	mg/L
$\mathbf{x}_{ieB}$	inert bulk liquid *E. coli* concentration	mg/L
$\mathbf{x}_{ipB}$	inert bulk liquid *P. putida* concentration	mg/L
$\mathbf{x}_{ef}$	viable *E. coli* attached cell concentration	g/m^3
$\mathbf{x}_{pf}$	viable *P. putida* attached cell concentration	g/m^3
$\mathbf{x}_{ief}$	inert *E. coli* attached cell concentration	g/m^3
$\mathbf{x}_{ipf}$	inert *P. putida* attached cell concentration	g/m^3

* the even variables are C, U, P, X_{aE}, X_{ap}, X_{iE}, and X_{ip}. However, mass balances for inert cells X_{iE} and X_{ip} may be lumped into one variable X_i to further simplify the model.

Biofilm Kinetics

The model is defined conceptually in terms of dissolved species (**u**) and biomass (**x**) as functions of time and space as shown in Figure 2. Mass balances for dissolved species and biomass across the entire reactor (Figure 2a) yield the following general equations:

$$\frac{\partial(\mathbf{u}_{pB}V_B)}{\partial t} = Q(\mathbf{u}_{in} - \mathbf{u}_{pB}) - \mathbf{r}_{uB} \cdot V_B - \mathbf{j}_u \cdot A_f \tag{1}$$

$$\frac{\partial(\mathbf{x}_{aB}V_B)}{\partial t} = -Q\mathbf{x}_{aB} + \lambda_a \cdot \mathbf{x}_{af} L_f A_f - b_x \mathbf{x}_{aB} V_B \tag{2}$$

$$\frac{\partial(\mathbf{x}_{iB}V_B)}{\partial t} = -Q\mathbf{x}_{iB} + \lambda_i \cdot \mathbf{x}_{if} L_f A_f \tag{3}$$

where $\mathbf{j}_u$ = flux of dissolved species into the biofilm ($ML^{-2}T^{-1}$), $\mathbf{r}_{uB}$ = rate of removal of dissolved species by suspended cells in the bulk liquid ($ML^{-3}T^{-1}$) (Appendix I), b_x = cell death rate coefficient (T^{-1}), λ_a and λ_i = viable and inert cell detachment rate coefficients (T^{-1}), Q = influent feed flow rate (L^3T^{-1}), and V_B = the bulk liquid volume (L^3).

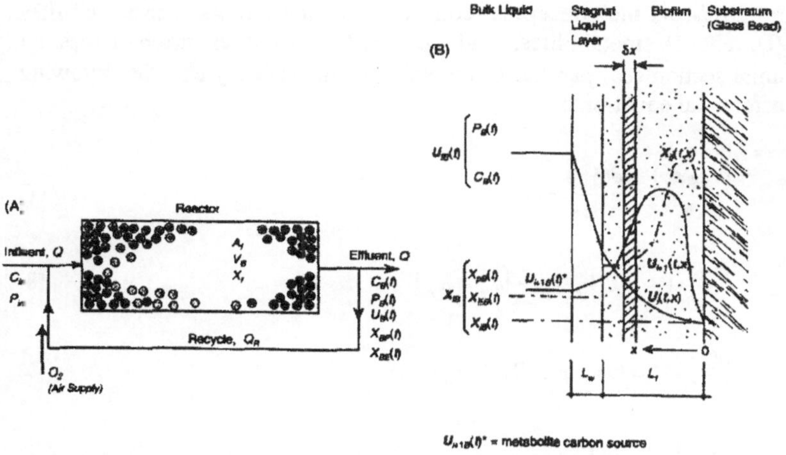

Figure 2. Conceptual biofilm and bioreactor model.

Mass transport of dissolved species obeys *Fick's* 2nd law of molecular diffusion:

$$\frac{\partial C_i}{\partial t} = D_{iw} \frac{\partial^2 C_i}{\partial z^2} \tag{4}$$

where C_i = concentration of dissolved species (ML^{-3}), D_{iw} = diffusivity of dissolved species in water (L^2T^{-1}), and z is the spatial coordinate (L). The diffusivities of phenol and metabolites were derived from Wilke and Chang (1956) [*10*]:

$$D_{iw} = \frac{(D'_i)(\phi M_B)^{0.5} T}{\mu \upsilon_A^{0.6}} \tag{5}$$

whereas the diffusivity of Cr(VI) was derived from the *Nernst-Haskell* equation:

$$D_{cw} = \frac{RT}{F^2} \cdot \frac{\lambda}{\gamma_i} \tag{6}$$

where D_{iw} = diffusion coefficient of species i at an absolute temperature T (°K), D_i' = diffusion coefficient of species i at standard temperature and pressure, ϕ = association factor for solvent (2.6 for water, M_B = molecular weight of solvent (g), ν_A = molar volume of solute (cm^3/g mole), and μ = solution viscosity, centipoise. For Cr(VI), a value of $D_{cw} = 1.05 \times 10^{-5}$ cm^2/s was found using a specific conductance $K = 4.5 \times 10^{-6}$ ohms/cm computed from tabulated values of $\Lambda = 1000 \cdot K/C$ for 1 M CrO_4^{2-} standard concentrations (25° C) [*11*].

Figure 2b shows the conceptual concentration and biomass density profiles of Cr(VI), phenol, metabolites, and cells. Mass balances taken across an infinitesimal section (δz) parallel to the substratum surface yields the following partial differential equations:

$$\frac{\partial(\varepsilon \cdot \mathbf{u}_p)}{\partial t} = \frac{\partial(\mathbf{j}_{up})}{\partial z} - \mathbf{r}_{fp} \tag{7}$$

$$\frac{\partial(\varepsilon \cdot \mathbf{x}_{pf})}{\partial t} = 1/Y_p \cdot \mathbf{r}_{pf} - (b_{xp} + \lambda_p) \cdot \mathbf{x}_{pf} \tag{8}$$

$$\frac{\partial(\varepsilon \cdot \mathbf{u}_u)}{\partial t} = 1/Y_u \cdot \mathbf{r}_{pf} - 1/Y_e \cdot \mathbf{r}_{ef}(\mathbf{x}_{ef}) \tag{9}$$

$$\frac{\partial(\varepsilon \cdot \mathbf{x}_{ef})}{\partial t} = \frac{\partial(\mathbf{j}_{xe})}{\partial z} + 1/Y_e \cdot (\mathbf{r}_{ef} - \Delta c) - (b_{xe} + \lambda_e) \cdot \mathbf{x}_{ef} \tag{10}$$

$$\frac{\partial(\varepsilon \cdot \mathbf{x}_{if})}{\partial t} = 1/Y_e \cdot \Delta c + (1 - f_d) \cdot b_{xe} \cdot \mathbf{x}_{ef} - \lambda \cdot x_{if} \tag{11}$$

$$\frac{\partial(\varepsilon \cdot \mathbf{u}_c)}{\partial t} = \frac{\partial(\mathbf{j}_c)}{\partial z} - \mathbf{r}_{ef} \tag{12}$$

with BCs:

$$\mathbf{j}_u = k_{Lu} \cdot (\mathbf{u}_B(t) - \mathbf{u}_{fs}(t, L_f)) \cdot A_f, \qquad z = L_f \tag{13}$$

$$\mathbf{j}_u = 0, \qquad z = 0 \tag{14}$$

where ε = is a biofilm porosity constant (V_{ftotal}/V_{fvoids}), k_{Lu} is the mass transfer rate coefficient, D_{uw}/L_w (L^2T^1), and $\mathbf{u}_{fs}$ = concentration of dissolved species at the liquid biofilm interface (ML^{-3}). At the observed low bulk liquid viable suspended biomass concentrations in this experiment, the rate of removal of dissolved species in the bulk liquid was insignificant ($\mathbf{r}_B \Rightarrow 0$). The kinetic parameters used in Equations 1-8 are summarized in Table II.

The model was solved by a numerical routine for partial differential equations written in the C/C++ and Fortran 77 languages. The model was solved in the temporal direction by a (fourth-order) Runge-Kutta method with spatial

TABLE II. KINETIC PARAMETERS FOR THE COCULTURE BIOFILM SYSTEM

Symbol		Units
k_{mc}	max. specific Cr(VI) reduction rate coeff.	1/h
K_c	Cr(VI) half velocity concentration	mg/L
R_c	*E. coli* cell Cr(VI) reduction capacity	mg/mg
b_{xp}	*P. putida* cell death rate coefficient	1/h
k_{mp}	max. specific phenol degradation rate coeff.	1/h
K_p	phenol half velocity concentration	mg/L
Y_p	*P. putida* cell yield coeff. from phenol	mg/mg
Y_e	*E. coli* cell yield coefficient from metabolites	mg/mg
b_{xe}	*E. coli* death rate coefficient	1/h
k_{mu}	max. specific metabolites utilization rate coeff.	1/h
K_u	metabolites half velocity concentration	mg/L
Y_u	metabolites yield coeff. from phenol	mg/mg
P_r	phenol inhibition threshold concentration	mg/L
K_{pe}	phenol inhibition coefficient on *E. coli* cells	mg/L
K_{io}	oxygen inhibition coeff. on Cr(VI) reduction	mg/L
K_{ip}	substrate inhibition coefficient - phenol	mg/L
C_r	Cr(VI) inhition threshold concentration	mg/L
K_{cp}	Cr(VI) inhibition coeff. on phenol degradation	mg/L
Input dynamic parameters		
ε	biofilm void ratio	--
f_d	biodegradable fraction of cells	--
λ_p	*P. putida* cell detachment rate coefficient	1/d
λ_e	*E coli* cell detachment rate coefficient	1/d
λ_i	Inert cell detachment rate coefficient	1/d
D_{cw}	Cr(VI) diffusion coefficient in water	cm^2/s
D_{pw}	phenol diffusion coefficient in water	cm^2/s
D_{uw}	metabolites diffusion coefficient in water	cm^2/s

discretization by the (second-order) Crank-Nicholson and Backward Euler finite difference methods.

RESULTS AND DISCUSSION

Transient-State Model Simulation

The biofilm mass balance equations (Equations 7 to 12) were coupled to reactor mass balances (Equations 1 to 3) through the outer boundary conditions (Equations 13 to 14) followed by solution of the biofilm system by finite difference methods. Effluent Cr(VI) and phenol concentration was used in the determination of kinetic parameters. Both Cr(VI) and phenol concentrations were determined with a high degree of certainty by colorimetric and gas chromatographic methods, respectively. Due to a relatively high diffusion coefficient of oxygen, DO was assumed non-limiting to the Cr(VI) reduction and phenol degradation processes in the reactor. DO kinetic terms were thus ignored during model development.

Initial Conditions

Viable cells in the biofilm zone were initialized to non-zero values X_{pf}ini = X_{ef}ini = 30 g/m^3 (established by trial-and-error). Bulk liquid concentrations of *P. putida* and *E. coli* cells (X_{pB}ini and X_{EB}ini) were set to 300 and 200 mg/L, respectively, representing the cell concentrations after inoculation at t = -30 days. Simulation was started at -30 days to represent the actual 30 day startup period in the reactor. Dissolved species concentrations in the bulk liquid and biofilm zones were initialized to zero ($P_{ini} = C_{ini} = U_{ini} = C_{fini} = P_{fini} = U_{fini} = 0$ mg/L) at t = -30 days, representing no loading before -30 days.

Comparison of Parameters Optimized at 24 and 6 hours HRT

The model was first optimized at 24 hours HRT (Phases I-VI, Figure 1) followed by optimization at 6 hours HRT (Phases XI-XIII), Figure 2). Results obtained from optimizations at the two hydraulic retention times were compared. Table III shows insignificant changes in parameter values due to optimization at different hydraulic times were ($\Delta q \leq 5\%$), for any given parameter (q) (Table III). Slight increases in the maximum reaction rate coefficients, k_{mc} and k_{mp}, after optimization at 6 hours HRT were attributed to culture acclimation at longer periods of operation. Higher cell detachment rate coefficients were expected at 6 hours HRT due to increases in shear forces on the biofilm. The best fitting parameters utilized a lump cell detachment rate coefficient, $\lambda_L = \lambda_p + \lambda_e + \lambda_i$, calibrated as an empirical function, $\lambda_L = \alpha_o + \beta_u \cdot u + \gamma \cdot (L_f)$. Changes in cell detachment rate coefficients in the range of 21-66% were observed after re-optimization at 6 hours HRT. Other operational and mass transport parameters such as the biofilm porosity factor (ε), and diffusion coefficients of dissolved species in water (D_{cw}, D_{pw}, and D_{uw}) were treated as inputs to the model.

Parameter Sensitivity

The transient state model was highly sensitive to the kinetic parameters (k_{mc}, R_c, k_{mp}, and K_p. Results from parameter analysis (not shown) demonstrated that changes in certain sets of parameters compensated each other during parameter optimization. For example, a positive change in k_{mc} may be compensated by a corresponding negative change in K_c. Thus, the results showed that there exists an infinite number of optima for any chosen set of kinetic parameters. False optima were rejected by locating the minimum combined MRSS for Cr(VI) reduction and substrate utilization rate coefficients

TABLE III. EFFECT OF HYDRAULIC LOADING ON OPTIMUM PARAMETERS.

Parameter *Symbol*	Values from Phases I-VI	Values from Phases XI-XIV	% Change in Parameter	Units
k_{mc}	0.094	0.095	1.1	1/h
K_c	10.2	9.1	-10.8	mg/L
R_c	0.12	0.13	8.3	mg/mg
b_{xp}	0.020	0.014	-30	1/h
k_{mp}	0.104	0.104	0	1/h
K_p	76.0	76.0	0	mg/L
Y_p	4.5	4.3	-4.4	mg/mg
b_{xe}	0.010	0.011	10	1/h
k_{mu}	0.035	0.035	0	1/h
K_u	25.4	25.3	-0.4	mg/L
Y_u	0.85	0.78	-8.2	mg/mg
Y_e	1.10	1.15	4.5	mg/mg
Input Paramers				
P_r	560	560	....	mg/L
K_{pe}	490	490	....	mg/L
K_{io}	9.1	9.1	....	mg/L
K_{ip}	79.0	281.0	....	mg/L
C_r	0.12	0.12	....	mg/L
K_{cp}	54	56	....	mg/L
λ_p	0.015	0.025	66.0	1/d
λ_e	0.007	0.0085	21.0	1/d
λ_i	0.446 [b]	0.586	31.2	1/d

[a] assumes insignificant detachment of viable biomass.
[b] cell detachment rate coefficient based on measured total biomass.
[c] operational parameters ε, f_d, D_{cw}, D_{pw}, and D_{uw} did not change.

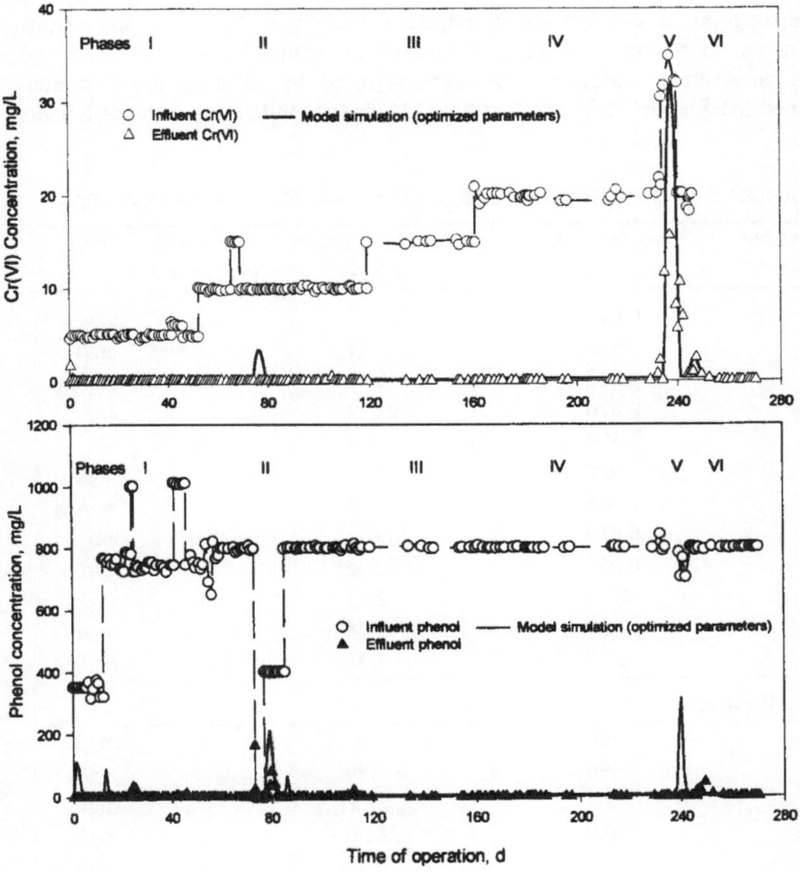

Figure 3. Parameter Optimization and Model Effluent Simulation at 24 hours HRT

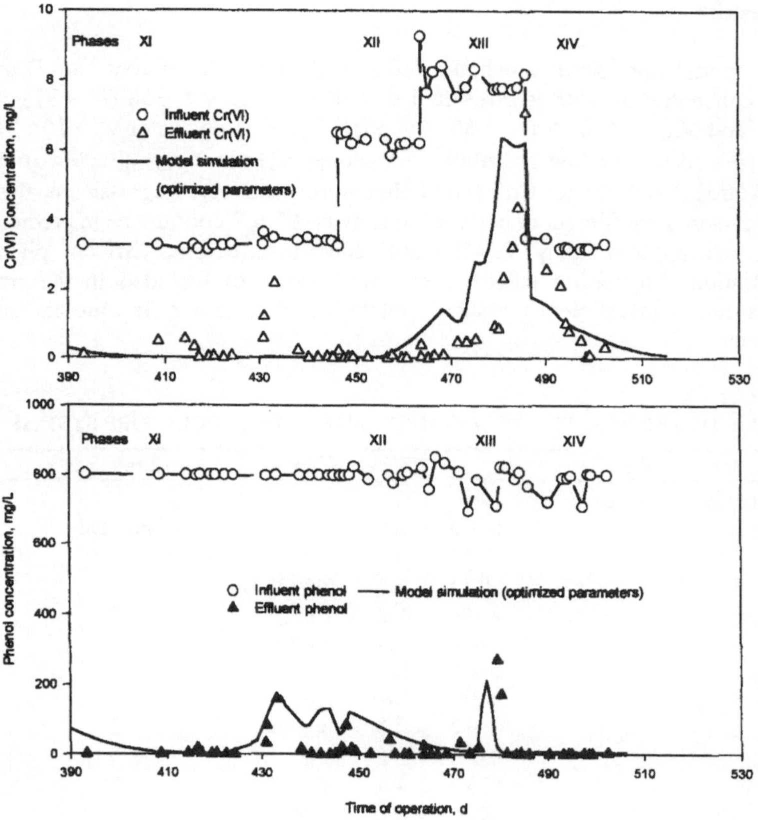

Figure 4. Parameter Optimization and Model Effluent Simulation at 6 hours HRT

during model optimization. Parameter sensitivity envelops were plotted from MRSS values were computed at 1% intervals over a grid of k_{mc}/K_c and R_c $(n\times m)$ = (5×5). The contours were generated using a 3D contour feature of Axum 5.0 (MathSoft Ware, Cambridge, UK). The parameter optimization process was later automated by using a heuristic approach (genetic search algorithm) adopted from Hunter (1994).

Best Results

The model optimized at 6 hours HRT predicted well effluent Cr(VI) and phenol concentration with low residual sum of squares, **V** = 358 (N = 97), for Cr(VI), and $\mathbf{V}_{(w)}$ = 1180.5 (N = 88) for phenol (weighing factor w = 10). The model yielded a very low probability for model rejection: $p_{\alpha(0.5)}(0.00086)$ and $p_{\alpha(0.5)}(0.00031) < 0.001$ for Cr(VI) and phenol, respectively. Regression analysis using Pearson's coefficient of correlation showed 98.6% confidence in predicted Cr(VI) and approximately 93.4% confidence in predicted effluent phenol concentration. Metabolite kinetic parameters were not included in the error analysis due to insufficient measured metabolites data under the transient-state conditions.

TABLE IV. THE BEST FITTING PARAMETERS FOR THE COCULTURE SYSTEM

Parameter	Value	CV%	Units	References
a. Optimized parameters				
k_{mc}	0.095	15	1/h	This study
K_c	9.1	8.2	mg/L	"
R_c	0.13	4.5	mg/mg	"
k_{mp}	0.104	0.87	1/h	"
K_p	76.0	7.7	mg/L	"
b_{xp}	0.014	1.80	1/h	"
b_{xe}	0.011	1.87	1/h	"
Y_p	4.3	5.8	mg/mg	"
k_{mu}	0.035	11	1/h	"
K_u	25.3	1.4	mg/L	"
Y_u	0.85	15.2	mg/mg	"
Y_e	1.15	15.0	mg/mg	"
b. Inputs				
$f_{dp}.f_{de}$	0.8	...		This study
P_r	560	8	mg/L	Shen and Wang, 1995b
K_{pe}	490	18.5	mg/L	"
K_{io}	10.3	14.2	mg/L	"
K_{ip}	79	16.7	mg/L	"
C_r	0.12	13.4	mg/L	"
K_{cp}	54	31.6	mg/L	"
α_0	0.442	9.4	1/d	This study
β_u	0.00105	9.5	1/μm	"
γ	0.004	32.4	1/μm/d	"

TABLE IV. CONTINUED

Parameter	Value	CV%	Units	References
D_{cw}	1.05×10^{-5}	3	cm^2/s	This study
D_{pw}	0.91×10^{-5}	25	cm^2/s	Wilke and Chen, 1966
D_{uw}	0.98×10^{-5}	25	cm^2/s	"
c. Operational parameters				
L_w†	40, 34, 25.4, 38.5	....	μm	Smets *et al.*, 1999
A_f	0.3140	....	m^2	This study
$Q_{in(1)}$	0.0001047	....	m^3/d	"
$Q_R/Q_{in(1)}$	300	....	m^3/d	"
V_B	0.0001047	....	m^3	"

† Stagnant liquid layer thickness determined by inspection at 24, 12, 6, and 18 hours HRT.

CONCLUSION

A rigorous biofilm model was developed to simulate Cr(VI) reduction, phenol and metabolites utilization, and biofilm dynamics and to optimize kinetic parameters in a *P. putida* DMP-1 and *E. coli* ATCC 33456 coculture biofilm reactor over a period of 687 days. Model predictions were consistent with experimental data for Cr(VI), phenol, and biomass concentration in the reactor. The model predicted effluent Cr(VI) concentration with 98.6% confidence in and effluent phenol with 93.4% confidence over various Cr(VI) loading conditions (2.5 to 32.5 mg/L/d).

ACKNOWLEDGMENTS

The research was partially funded by the United States Environmental Protection Agency through an exploratory research grant awarded to Yi-Tin Wang under the agreement R-822307-01-0. Special thanks are due to Seongjai Kim of University of Kentucky Mathematics Department for numerical simulations and PDE algorithms.

APPENDIX I. Reaction Rate Laws

The rate Cr(VI) reduction and phenol degradation were derived by Shen and Wang (1995b) from batch kinetics [*4*]:

$$r_c = \frac{k_{mc} \cdot C \cdot X_e^o}{K_c + C \cdot (1 + O_2/K_{io})} \cdot \left(\frac{1}{1 + (P - P_r)/K_{pe}} - \frac{C_o - C}{X_e^o R_c} \right) \qquad (15)$$

$$r_p = \frac{k_{mp} \cdot P}{K_p + P + P^2/K_{ip}} \cdot \frac{X_p^o}{1 + (C - C_r)/K_{cp}} \qquad (16)$$

REFERENCES

1. Kuo, C. W., and Genthner, B. R. S. 1996. "Effect of added heavy metal ions on biotransformation and biodegradation of 2-chlorophenol and 3-chlorobenzoate in anaerobic bacterial consortia." *App. & Environ. Microb.*, 62(7): 2317-2323.

2. Shen, H., and Y. T. Wang. 1995. "Simultaneous chromium reduction and phenol degradation in a coculture of *Escherichia coli* ATCC 33456 and *Pseudomonas putida* DMP-1." *App. & Environ. Microb.*, 61(7): 2754-2758.

3. Nkhalambayausi-Chirwa, E., and Wang, Y. T. 2000. "Simultaneous Chromium(VI) Reduction and Phenol Degradation in a Fixed-Film Coculture Bioreactor: Reactor Performance." *Water Research*, 35(8): 1921-1932.

4. Shen, H., and Wang, Y. T. 1995b. "Modeling simultaneous hexavalent chromium reduction and phenol degradation by a defined coculture of bacteria." *Biotechnology and Bioengineering*, 48(6): 606-613.

5. Wang, Y. T., and M. Qu. 1992. "Substrate interactions during biodegradation of phenols by a *Pseudomonas* sp." *Proceedings of the 65th Annual WEFTEC*, Alexandria, VA, pp. 63, October 5-9, 1996.

6. American Public Health Association. 1992. Standard Methods for the Examination of Water and Wastewater, 19th Edition.

7. Hollender, J., J. Hopp, and W. Dott. 1997. "Degradation of 4-chlorophenol via the meta cleavage pathway by *Comamonas testosteroni* JH5." *App. & Environ. Microb.*, 63(11): 4567-4572.

8. Chirwa, E. M. N., and Y. T. Wang. 1997. "Chromium(VI) reduction by *Pseudomonas fluorescens* LB300 in fixed-film bioreactor." *Journal of Environmental Engineering*, 123(8): 760-766.

9. Caldwell, D. E., and Lawrence, J. R. 1986. "Growth kinetics of *Pseudomonas fluorescens* microcolonies within the hydrodynamic boundary layers of surface microenvironments." *Microbial Ecology*, 12: 299-312.

10. Wilke, C. R., and Chang, P. 1955. "Correlation of diffusion coefficients in dilute solutions." *AIChE Journal*, 1:264.

11. Perry, R. H. 1997. Perry's Chemical Engineers Handbook. 7th Edition, Eds. Perry, R. H., Green, D. W., and Maloney, J. O., McGraw-Hill, New York, NY.

ANAEROBIC TRANSFORMATION AND BIOREMEDIATION OF CEPHALEXIN ANTIBIOTIC WASTE SLUDGE USING A FLUIDIZED BED REACTOR

R. Saravanane*, D. V. S. Murthy and K. Krishnaiah

Environmental Engineering Laboratory, Department of Chemical Engineering, Indian Institute of Technology, Madras, Chennai - 600 036, India

Abstract- Cephalexin is a semisynthetic cephalosporin antibiotic intended for oral administration. The nucleus of cephalexin is related to that of other cephalosporin antibiotics. The effluent and sludge from these industries contain high concentration of cephalexin and its intermediates which results in depletion of oxygen and destruction of aquatic animals due to their toxic nature. Anaerobic transformation of cephralexin occurred through *Pseudomonas aeroginosa*, *Escherichia coli* and *Achromobacter* species available in the mixed culture of an anaerobic digester. Cephralexin was transformed to 7-amino-3-deacetoxycephalosporanic acid (7-ADCA) and acyl group by the microorganisms in the presence of volatile fatty acid as co-substrate. This study was conducted to examine the feasibility of using an anaerobic fluidized bed reactor for the transformation of cephalexin. Rapid start-up of the reactor was achieved by using anaerobic cephalosporin enrichment cultures as the inoculum. During the continuous operating periods, low concentrations (3.5 mg/L) of cephalexin were transformed to 7-ADCA and acyl group at hydraulic retention times of 15-48 hours with 150 mg COD/L of volatile fatty acid. Cephalexin concentrations as high as 80 mg/L was transformed to 7-ADCA with a relatively low supply of (200 mg/L) of volatile fatty acid. Results of this study suggest that the anaerobic fluidized bed system is a feasible process for the anaerobic treatment and remediation of cephalexin based pharmaceutical waste sludge and effluent.

Key words- Anaerobic transformation, cephalexin, fluidized bed reactor, pharmaceutical effluent, bioaugmentation, enricher reactor

Corresponding author: **R. SARAVANANE,** Environmental Engineering Laboratory, Department of Chemical Engineering, Indian Institute of Technology, Madras, Chennai - 600 036, India
E-mail: perusara@yahoo.com; Fax: 91-44-2350305

EFFECTS OF NICKEL ON NITRIFYING ENRICHMENT CULTURES

ZHIQIANG HU, KARTIK CHANDRAN, BARTH F. SMETS
Environmental Engineering Program, Department of Civil and Environmental Engineering, University of Connecticut, Storrs, CT 06269

DOMENICO GRASSO
Picker Engineering Program, Smith College, Northampton, MA 01063

INTRODUCTION

Nitrification involves the conversion of ammonium-nitrogen (NH_4^+-N) to nitrite (NO_2^--N) and nitrate (NO_3^--N). It is generally considered the rate limiting step in biological nitrogen removal (1), because of the low biomass yield of the nitrifying bacteria and their extreme sensitivity to a number of environmental conditions such as pH, dissolved oxygen concentration, temperature (2-5).

A variety of chemical inhibitors can also adversely affect the specific growth rate of the nitrifying bacteria, resulting in either the failure or reduction of nitrification (6). Heavy metals, often derived from industrial wastewater contributions, are one of the causative agents to nitrification inhibition in wastewater treatment plants (WWTPs)(7, 8). The mechanism of heavy metal toxicity to nitrifying enrichment cultures is not clear. Yet in general, it is thought that heavy metal toxicity to microorganism depends on the metal cations entering the cell via nonspecific metal inorganic transport (MIT) (9). Once inside the cell, heavy metals tend to bind to thiol groups and inhibit enzyme activity. In addition, heavy metals may compete with physiological ions, e.g., Cd^{2+} with Zn^{2+} or Ca^{2+}, Ni^{2+} with Fe^{2+}, Zn^{2+} with Mg^{2+}, and thereby inhibiting the function of the respective physiological cations (9).

The toxicity of heavy metals to microorganisms is dependent on their bioavailability, defined here as the amount of a given heavy metal that produces an observed change in enzymatic activity. If mass transfer by physical transport (i.e., hydrodynamic mixing) and chemical transport (i.e., chemical reaction) are assumed to be more rapid than biological uptake, which is the general case, pseudo-equilibrium is established between the metal in the bulk solution and that associated with the surface of microorganism (10). Then, as depicted in the free ion activity model (FIAM) (11), heavy metal toxicity is directly proportional to the activity of the free ion in solution. Metal complexation with ligands in bulk solution and many functional groups on the microbial surface can decrease the activity of the free ion in solution, and thus attenuate the heavy metal toxicity (12-14).

Metal complexes are categorized into two groups, labile and non-labile (inert) (15). Metal lability in biouptake systems satisfies the following two conditions: (1) the non-dimensional pseudo first order forward and backward rate constants are far higher than unity; (2) supply of metal through diffusion toward microbial surface is not limited by the association/dissociation kinetics of the complex species (15, 16). Non-labile metal complex species like Ni-EDTA are generally considered non-bioavailable (17) because of their very slow dissociation rate. On the contrary, labile metal species may be considered bioavailable (16).

The primary objectives of this work were to determine and quantify the toxicity of nickel to a nitrifying enrichment culture and determine the effect of nickel species on toxicity. Nickel was investigated because Ni alloys are widely used in industry (18) and Ni exists mainly in soluble form in wastewater plants (19, 20).

MATERIALS AND METHODS

Nitrifying enrichment culture. Nitrifying enrichment culture was grown in a continuous reactor running at solids retention time (SRT) of 20 d and hydraulic retention time (HRT) of 1 d. The reactor had a total volume of 10 L. An inorganic medium devoid of organic carbon was fed to the reactor. Ammonia sulfate (300 mg/L N) was provided as the sole energy source in a feeding solution containing the requisite macro- and micronutrients (Ni background concentration 0.07 mg/L). Sodium carbonate (1 *M*) was periodically added to the reactor to fulfill both carbon and alkalinity requirements. The reactor was operated in the pH range of 7.4± 0.1. Filtered laboratory air was provided to ensure adequate mixing and aeration. Reactor performance was monitored via reactor and effluent COD, effluent NH_4^+-N, NO_2^--N, and NO_3^--N concentrations. Upon attainment of steady state, mixed liquor was periodically withdrawn from reactor and used for batch studies.

Batch respirometric assay for monitoring nitrification activity. Maximum specific oxygen uptake rates were measured using the batch respirometric assay described earlier (21). Briefly, aliquots (50 mL) of the nitrifying enrichment cultures were collected from the bench-scale reactor at pH 7.4 ± 0.05. Adequate buffer capacity was provided by adding 20 m*M* 3-(*N*-Morpholino)propanesulphonic acid (MOPS, pH adjusted to 7) in the cultures and the pH was adjusted to 7.0± 0.05 finally to avoid the significant complexation of metal with hydroxyl ion at pH 7.5. The assay was performed in a nominal 50 mL jacketed glass vessel, maintained at 25± 0.5°C. Substrate (NH_4^+-N or NO_2^--N) was injected in the biomass samples super-saturated with pure oxygen. A decrease in the dissolved oxygen (DO) level in the vessel due to substrate oxidation was measured.

Nitrification inhibition was investigated in the absence or presence of specifically added metal complexing agents. Metal complexing agents such as sulfate, citrate, nitrilotriacetic acid (NTA), and ehtylenediaminetetraacetate (EDTA) with their Ni formation constant range of $10^{2.3}$ to $10^{20.4}$ were chosen. Aliquots of metal complexing agents were mixed with aliquots of Ni stock solution to give final concentrations of metal complexing agents and Ni of 0-10 m*M*, 20-60 mg/L respectively. Ni^{2+} concentration was thus manipulated by maintaining constant Ni concentration while changing the concentrations of metal complexing agent or by maintaining constant concentrations of metal complexing agents while varying the Ni concentration in the cultures. The concentrated Ni or Ni complexes were then spiked into batch respirometric vessels. Thereafter, the standard procedure for the batch respirometric assay was followed.

Nitrification inhibition was calculated from the difference of measured SOUR in the absence($SOUR_{control}$) and presence ($SOUR_{sample}$) of toxic treatment according to following formula:

$$\% \text{ Inhibition} = (SOUR_{control} - SOUR_{sample}) / SOUR_{control} * 100\% \quad (1)$$

Differentiation of soluble Ni and intracellular Ni. At the end of the batch respirometric assay, aliquots (30 mL) of the nitrifying enrichment cultures from the jacketed glass vessel were collected in 50 mL nominal polyethylene centrifugation tubes and centrifuged at 16,000 g (Sorvall, Model RC-5C Plus) for 10 minutes. The supernatants were carefully removed using 5 mL pipette and were passed through 0.45 μm membrane (Millipore, MA). 10 mL filtrate were used for soluble Ni (< 0.45 μm) measurement. The remaining filtrate was further passed through 1,000 molecular weight cut-off (MWCO) ultrafiltration membrane (YM series, Millipore, MA) to determine the truly soluble Ni. Ultrafiltration was conducted in 50-mL stirred ultrafiltration cells (Amicon, Model 8050) using He at 345 KPa. Ultrafiltration membranes were initially conditioned according to manufacturer's instructions.

Intracellular Ni was measured by an optimized EDTA washing procedure (10, 14, 22, 23). The pellets after the centrifugation were re-suspended in 30 mL 1 m*M* EDTA solution and placed on a shaker at 150 rpm for 30 minutes to remove Ni bound to the microbial surface. The suspensions were then centrifuged at 16,000g (x 2) and 2.5 mL concentrate nitric acid (trace metal grade) was added in the 50 mL nominal polyethylene centrifugation tubes containing the pellets. The solution was transferred into 10 mL nominal COD vials by sequential washing with 2.5 mL DI water (x 3). Finally, the samples were digested for 24 h at 100 °C. The digested suspension was filtered through 0.45 μm membrane and the filtration Ni was measured using atomic absorption spectrophotometer.

Metal species calculation. Ni speciation was calculated with the MINEQL+ chemical equilibrium model (24), with incorporation of complexation constant for Ni-MOPS and Ni-biomass.

MOPS was chosen as a buffer source because its maximum buffer capacity (pKa 7.2, 20°C) is around pH 7 in the batch respirometric assay. Given its sulfonic structure, a low metal complexation capacity would be expected. A conditional formation constant ($K=10^{2.3}$) between copper and MOPS was experimentally determined by measuring the change of free cupric ion as a function of total copper spiked in pH 7 MOPS solution. According to empirical Irving-Williams series (25), the stability of metal complexes follows the series

$$Mn^{2+} < Fe^{2+} < Co^{2+} < Ni^{2+} < Cu^{2+} > Zn^{2+}$$

As a result, a conservative conditional formation constant ($K=10^{2.3}$) of Ni-MOPS was applied in Ni species calculation.

Similarly, a conditional formation constant ($K=10^{2.0}$) of Ni with nitrifying cultures was calculated from the linear total-soluble Ni partitioning coefficient obtained in batch respirometric assay with further assumption that free Ni cation is proportional to the soluble Ni.

Ni organic complexation was insignificant in the species calculation because the feeding media was devoid of organics and the measured soluble organic compound in the reactor was below 20 mg/L COD (approximately 2 % of biomass concentration in the reactor). All other formation constants were used from MINEQL+.

Analytical procedures. All measurements are according to standard methods (26). NH_4^+-N was analyzed using an ammonia gas-sensing electrode (HNU Systems MA) with a range of operation between $7*10^{-3}$ to $1.4*10^4$ mg/L NH_4^+-N. NO_2^-- N was measured using a modification of the sulfanilic acid-N-naphthyethylenedamine addition method. NO_3^-- N was measured using ionic chromatography (Dionex, Model DX-500, CA). Nitrifying culture concentration was measured as COD using commercially available reagents (HACH COD vials, 0-1500 mg/L). Ni was measured by flame atomic absorption spectrophotometer (Model 5100, Perkin-Elmer Co., CT) with a detection limit of 0.04 mg/L Ni. Precautions were taken to avoid trace metal contamination. Glassware and centrifugation tubes were soaked in 1 *M* HNO3 overnight and rinsed with Milli-Q water five times before use.

RESULTS AND DISCUSSION

Effect of nickel and EDTA on rate of ammonium and nitrite oxidation. The specific oxygen uptake due to ammonium oxidation ($SOUR_{NH4}$) decreased as nickel concentration in the nitrifying enrichment cultures increased (Fig. 1). In contrast, the specific oxygen uptake due to nitrite oxidation ($SOUR_{NO2}$) was constant within the range of 100 mg/L Ni tested. This was in agreement with the literatures that ammonium oxidation is generally more sensitive to inhibitory compounds than that nitrite oxidation (6, 27).

Analogously, $SOUR_{NH4}$ decreased significantly when the EDTA concentration was higher than 1 m*M* in the nitrifying enrichment cultures (Fig.2). EDTA may complex with essential cations like calcium (Ca^{2+}), Magnesium (Mg^{2+}), copper (Cu^{2+}), zinc (Zn^{2+}), etc. inside the cell at concentration higher than 1 m*M*, leading to the inhibition to the enzyme activity. Below 1 m*M*, inhibition of ammonium oxidation by EDTA was not observed. As a result, 1 m*M* EDTA was used in the following Ni-EDTA toxicity and intracellular Ni measurement experiments. In addition, there was no effect of EDTA on nitrite oxidation at tested concentrations.

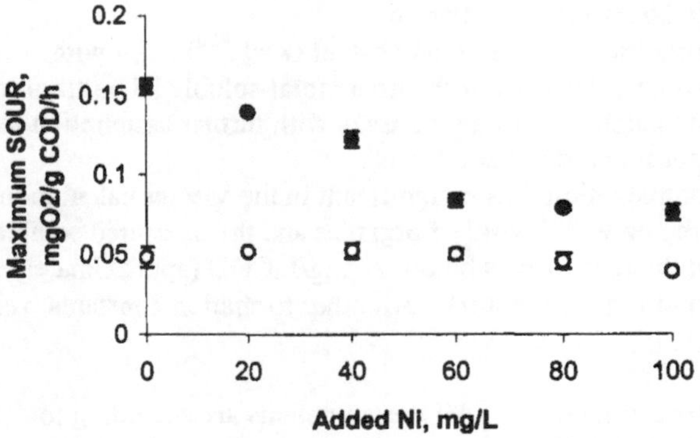

Figure 1. Effect of nickel on maximum specific oxygen uptake rates in nitrifying enrichment cultures. •, NH_4^+-N oxidation; o, NO_2^--N oxidation.

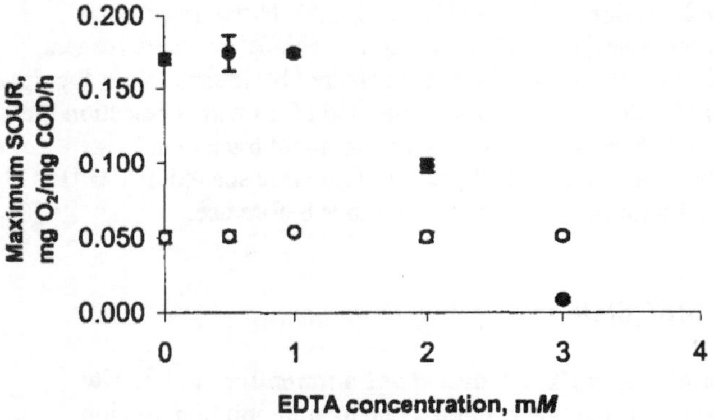

Figure 2. Effect of EDTA on maximum specific oxygen uptake rates in nitrifying enrichment cultures. •, NH_4^+-N oxidation; o, NO_2^--N oxidation.

Ni species and their toxicity in the absence of metal complexing agents. By plotting the measured Ni concentrations vs. the total Ni injected in the nitrifying enrichment cultures (Fig.3), a strong correlation between soluble Ni (< 0.45 μm portion) and total Ni was observed (R^2 =0.9999). Similarly, a strong correlation between truly soluble Ni (< 1,000 MWCO) and total Ni existed (R^2 =0.9963). Of the total Ni added, 87.5 % Ni was in soluble form (< 0.45 μm portion) while only 60.7 % Ni was in truly soluble form (<

1,000 MWCO). Linear correlation between intracellular Ni and total Ni in the cultures was also observed with lower relative coefficient (R^2 =0.8894).

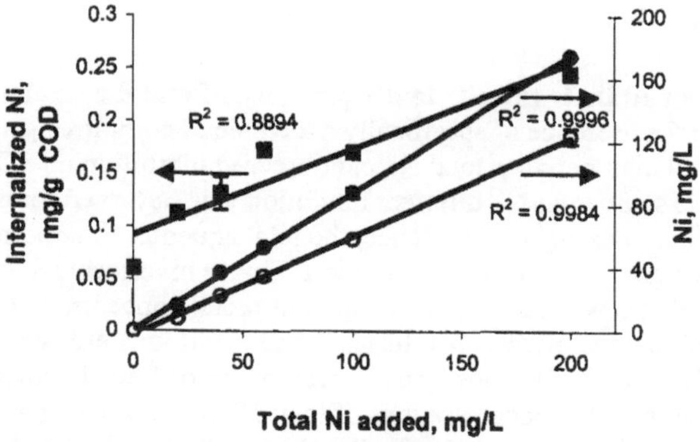

FIGURE 3. Analytical nickel concentrations as a function of the total Ni injected in nitrifying enrichment cultures. Symbols: ϖ, intracellular Ni; •, soluble Ni (< 0.45 μm); o, truly soluble Ni (<1,000 MWCO).

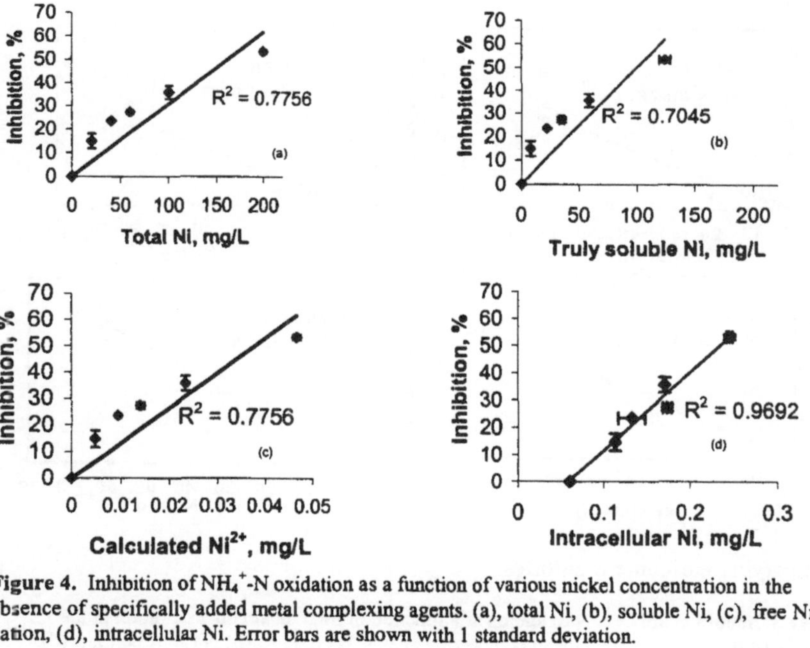

Figure 4. Inhibition of NH_4^+-N oxidation as a function of various nickel concentration in the absence of specifically added metal complexing agents. (a), total Ni, (b), soluble Ni, (c), free Ni cation, (d), intracellular Ni. Error bars are shown with 1 standard deviation.

Various Ni fractions were then correlated with Ni toxicity (inhibition of $SOUR_{NH4}$) in the absence of specifically added metal complexing agents (Fig 4a-d). The inhibition increased quickly until the total nickel spiked in culture reached 60 mg/L Ni, and then increased only slightly at higher total Ni concentration (maximum 53 % inhibition at 200 mg/L Ni) (Fig 4a). Similarly, there was no constant linear correlation between soluble Ni (Fig. 4b) or free nickel ion (Fig. 4c) and observed inhibition as well. Instead, inhibition was linearly related to the cell internalized (intracellular) nickel concentration (R^2 = 0.9692) (Fig. 4d).

Ni species and their toxicity in the presence of metal complexing agents. In the presence of specifically added metal complexing agents, there was no correlation between total Ni and observed inhibition on $SOUR_{NH4}$ (Fig.5a-d). For the same total Ni, different inhibition was observed for different concentration of complexing agents. Hence, the total aqueous Ni concentration is not a good predictor of nitrification inhibition. This is in agreement with documented reports showing that the total aqueous metal concentration is not a good predictor of its toxic effects (14). Instead, free metal ions are generally the most toxic metal species (28). If the same concentration of free Ni cation was maintained by changing the concentrations of total Ni or complex agents, a constant degree of inhibition would be predicted according to the free ion model (11). This was supported well by our results in Figure 6(a-d). We also observed that inhibition of NH_4^+-N oxidation increased as free Ni cation increased, indicating that free Ni cation is most likely responsible for the Ni toxic effects on nitrifying enrichment cultures.

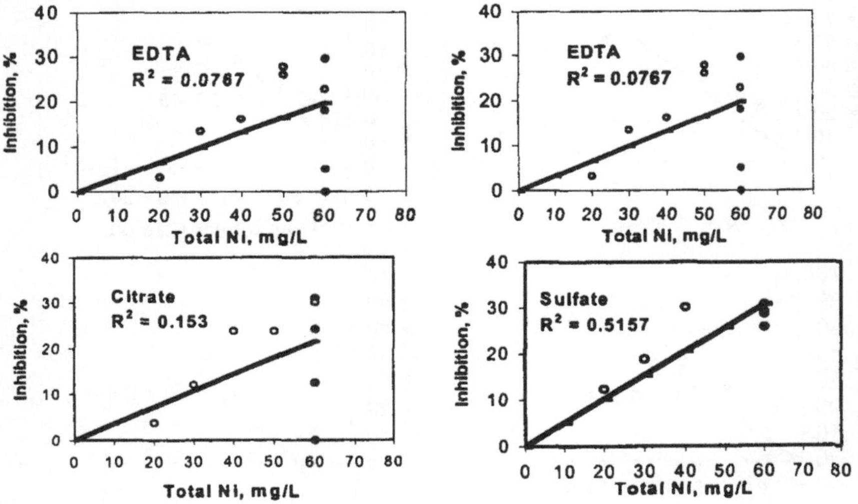

Figure 5. Inhibition of NH_4^+-N oxidation as a function of total nickel concentration in the presence of specifically added metal complexing agents. (a) EDTA, (b) NTA, (c) citrate, (d) sulfate. Symbols: •, free Ni cation manipulated by keeping the same total Ni concentration while varying the metal complexing agent concentrations; o, free Ni cation manipulated by keeping the same metal complexing agent concentration while varying the total Ni concentrations).

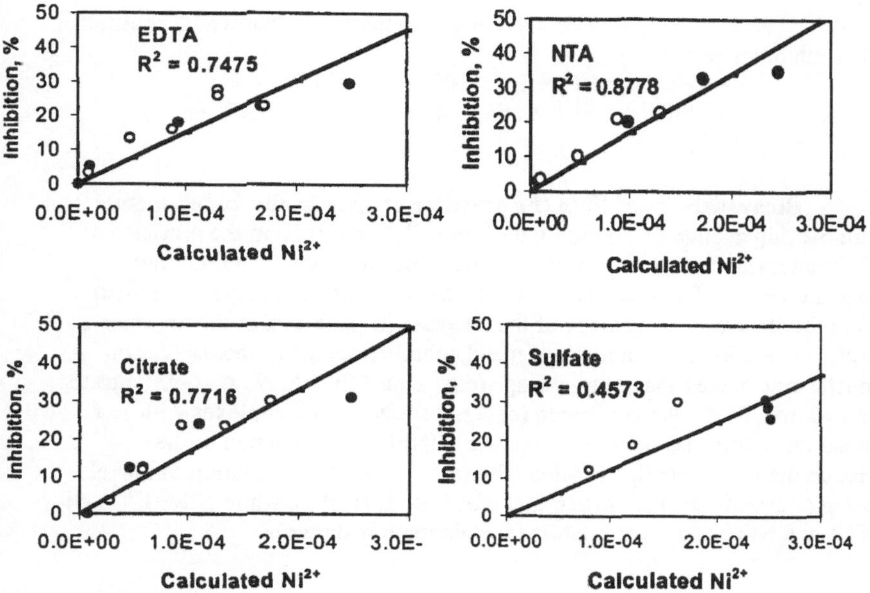

Figure 6. Inhibition of NH_4^+-N oxidation as a function of Ni^{2+} concentration in the presence of specifically added metal complexing agents. (a) EDTA, (b) NTA, (c) citrate, (d) sulfate. Symbols: •, free Ni cation manipulated by keeping the same total Ni concentration while varying the metal complexing agent concentrations; o, free Ni cation manipulated by keeping the same metal complexing agent concentration while varying the total Ni concentrations).

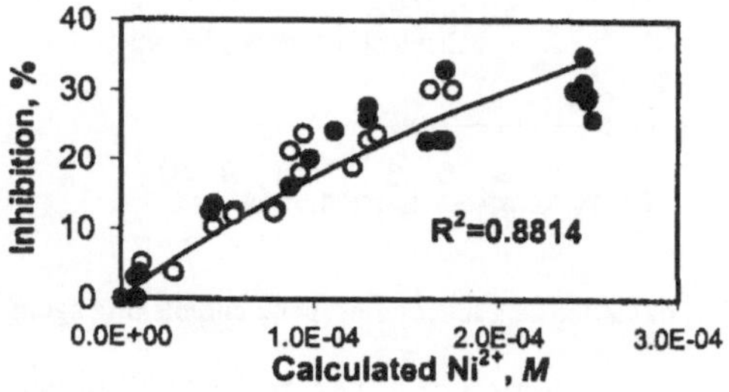

Figure 7. Inhibition of NH_4^+-N oxidation as a function of free nickel ion concentration. Data are all complied from the toxicity assay in the absence and presence of specifically added metal complexing agents. A non-competitive inhibition model predicted line is shown as well.

Combining all the data from the experiments of Ni toxicity in the presence of metal complexing agents, the inhibition on $SOUR_{NH4}$ increases as the predicted Ni concentration increases with a non-linear pattern (Fig. 6-7). An empirical non-competitive inhibition model using simplified equation 2 was used to best fit the full data set (R^2=0.8814) and gave an inhibitor coefficient K_i

of $4.76*10^{-4}$ *M*. K_i represents the inhibitor concentration that would result in 50% inhibition of $SOUR_{NH4}$.

$$i = 1/(1+ K_i/ I)*100 \qquad (2)$$

where I = inhibitor concentration; K_i = inhibition coefficient;
i = % of inhibition.

Bioavailability of Ni in the presence of specifically added metal complexing agents. Comparing the bioavailability of Ni in the presence of different metal complexing agents, the trend of inhibition on ammonium oxidation was different depending on the metal complexing agent as evident from the steepness of the slope of the decrease in inhibition in the presence of 60 mg/L Ni (Fig 8). No inhibition of metal complexing agents themselves on nitrification at their maximum dosage (e.g., 1 m*M* EDTA, NTA, 2 m*M* citrate, and 10 m*M* SO_4^{2-}) was confirmed (data not shown). Ni complexes with 1: 1 molar ratio were also observed from the MINEQL+ calculation as the predominated Ni complex species. The results from the evaluation of nickel complex lability showed further that only $NiSO_4$ is labile while $NiEDTA^{2-}$, Ni-NTA and Ni-citrate are non-labile (calculation not shown).

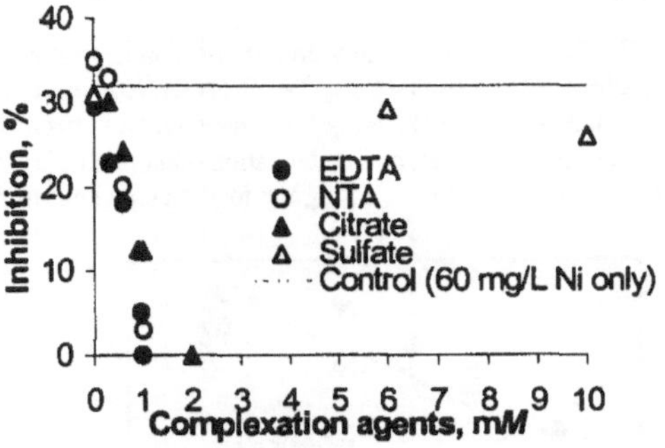

Figure 8. Inhibition of NH_4^+-N oxidation as a function of metal complexing agent concentration.

At 1 m*M* concentration, addition of EDTA reduced Ni bioavailability and the Ni toxicity was not observed; 100% of the spiked Ni existed as $NiEDTA^{2-}$. This was in agreement with documented literature that $NiEDTA^{2-}$ is essentially inert and non-bioavailable (17). The formation constants (K) of Ni-EDTA and Ni-NTA are $10^{20.4}$, and $10^{12.8}$, respectively (25). Similar inhibition pattern was applied for NTA as well, because free Ni cation was calculated based on equilibrium condition where extra EDTA could strongly complex Ca^{2+}, Mg^{2+} ions in the bulk solution.

Addition 1 m*M* citrate moderately reduced Ni bioavailability and Ni toxicity (12 % inhibition to NH_4^+-N oxidation) was still observed. As the citrate concentration increased, the calculated Ni-citrate complex ($K= 10^{6.7}$) concentration increased as well while the Ni toxicity was attenuated. At 2 m*M* citrate concentration, no Ni toxicity was observed.

Addition 1 m*M* sulfate basically had no effect on Ni bioavailability as indicated from the relative constant inhibition on $SOUR_{NH4}$ before and after sulfate addition. Even 10 m*M* sulfate concentration was added, Ni toxicity still persisted. As $NiSO_4$ (K= 2.3) is formed by weak electrostatic reaction, large amount of labile $NiSO_4$ species provide high concentration of free Ni cation, which may be directly utilized by the cell and causes nitrification inhibition.

In conclusion, inhibition of ammonium oxidation by Ni is primarily caused by free Ni cation, while the free Ni cation concentration is governed by the ligand concentration and affinity to free Ni cation. Inhibition of Ni on ammonium oxidation can be attenuated by adding appropriate strong ligand (EDTA) in the solution.

REFERENCES

1. U.S. EPA. 1993. Nitrogen control. In *EPA/625/R-93-0101*. Washington, D. C.
2. Dangcong, P., N. Bernet, J. Delgenes, and R. Moletta. 2000. *Water Environ. Res.* 72: 195-200.
3. Ford, D. L., R. L. Churchwell, and J. W. Kachtick. 1980. *J. Wat. Pollut. Control. Fed.* 52: 2726-36.
4. Shammas, N. K. 1986. *J. Wat. Pollut. Control. Fed.* 58: 52-8.
5. Painter, H. A. 1970. *Water Res.* 4: 393-450.
6. Blum, D. J. W., and R. E. Speece. 1991. *Research Journal of the Water Pollution Control Federation* 63: 198-207.
7. Stensel, H. D., C. S. McDowell, and E. D. Ritter. 1976. *J. Wat. Pollut. Control. Fed.* 48: 2343-50.
8. Gruttner, H., M. Winther-Nielsen, L. Jorgensen, P. Bogebjerg, and O. Sinkjer. 1994. *Water Science and Technology* 29: 69-77.
9. Nies, D. H. 1999. *Applied Microbial Biotechnol.* 51: 730-50.
10. Mirimanoff, N., and W. K. J. 2000. *Environ. Sci. Technol.* 34: 616-22.
11. Morel, F. M. M., and J. G. Hering. 1993. *Principles and applications of aquatic chemistry*. New York: John Wiley & Sons, INC. pp.405-14.
12. Yoshizaki, S., and T. Tomida. 2000. *Environ. Sci. Technol.* 34: 1572-5.
13. Morton, J. D., K. F. Hayes, and J. D. Semrau. 2000. *Environ. Sci. Technol.* 34: 4917-22.
14. Parent, L., M. R. Twiss, and P. G. C. Campbell. 1996. *Environ. Sci. Technol.* 30: 1713-20.
15. van Leeuwen, H. P. 1999. *Environ. Sci. Technol.* 33: 3743-8.
16. Jasen, R. A. G., H. P. van Leeuwen, R. F. M. J. Cleven, and v. D. H. M. A. G. T. 1998. *Environ. Sci. Technol.* 32: 3882-6.
17. Bedsworth, W. W., and D. L. Sedlak. 1999. *Environ. Sci. Technol.* 33: 926-31.
18. Bernner, A. 1963. *Principles and practice*. New York: Academic Press. pp.239-314.

19. Wang, J., C. P. Huang, H. E. Allen, I. Poesponegoro, H. Poesponegoro, and L. R. Takiyama. 1999. *Water Environ. Res.* 71: 139-47.
20. Monteith, H. D., J. P. Bell, and D. J. Thompson. 1993. *Water Environ. Res.* 65: 129-37.
21. Chandran, K., and B. F. Smets. 2000. *Biotechnol. Bioeng.* 68: 396-406.
22. Chang, S. I., and J. R. Reinfelder. 2000. *Environ. Sci. Technol.* 34: 4931-5.
23. Texier, A., Y. Andres, M. Illemassene, and P. L. Cloirec. 2000. *Environ. Sci. Technol.* 34: 610-5.
24. Schecher, W. D., and D. C. McAvoy. 1998. MINEQL+: a chemical equilibrium modeling system. Hallowell, ME: Environmental Research Software
25. Stumm, W., and J. J. Morgan. 1996. *Aquatic chemistry.* New York: John Wiley & Sons, INC.
26. APHA, AWWA, and WEF. 1995. *Standard methods for the examination of water and wastewater*
27. Tang, N. H., D. J. W. Blum, N. Nirmalakhandan, and R. E. Speece. 1992. *Journal of Environmental Engineering* 118: 17-37.
28. Shuttleworth, K. L., and R. F. Unz. 1991. *Water Res.* 25: 1177-86

Enhancing Anaerobic Dehalogenation of Organhalides in Contaminated Sediments

Max M. Häggblom [1], D.E. Fennell [1], L.J. Kerkhof [2], S.-K. Rhee [1,2] and Y.-B. Ahn [1]

Dept. of Biochemistry and Microbiology
and Biotechnology Center for Agriculture and the Environment [1]
and Institute of Marine and Coastal Sciences [2]
Cook College, Rutgers, The State University of New Jersey
New Brunswick, NJ 08901

Halogenated compounds constitute one of the largest groups of environmental pollutants, partly as a result of their widespread use as biocides, solvents and other industrial chemicals. Marine and estuarine sediments are significant sinks particularly for aromatic organohalide compounds and the management of these contaminated sediments is a major problem around the world. Microbial degradation of halogenated organic compounds in anoxic marine sediments generally proceeds first via reductive dehalogenation. This process can proceed simultaneously with other dominant electron-accepting processes, with further degradation of the dehalogenated product coupled to methanogenesis, sulfate reduction, and iron (III) reduction. The objective of our work has been to determine the capacity of microbial communities in anoxic sediments for dehalogenation and degradation of organohalides and to develop strategies to enhance their activities for in situ bioremediation. A detailed characterization of sediment microbial communities is needed to fully understand the role that different groups play in the degradation of haloaromatics under different conditions. We are identifying important phylogenetic and physiological groups and assessing their activity through ribosome and gene expression analysis. We are using redox manipulation and amendment of simple halogenated compounds to enhance the activity of dehalogenating bacteria that are responsible for contaminant dehalogenation. Evaluation of the roles of microbial community members under different conditions during organohalide dehalogenation will eventually aid in design of in situ techniques for sediment bioremediation.

SYMPOSIUM 4:

Integrated Management of Municipal Solid Waste

DO MODERN MSW INCINERATORS ACTUALLY DESTROY MORE PCDD/F THAN THEY PRODUCE?

Hsien-Tsung Chern
Dept. of Chemical Eng., Chemistry and Environmental Science,
New Jersey Institute of Technology, Newark, NJ 07102

Joseph W. Bozzelli
Dept. of Chemical Eng., Chemistry and Environmental Science,
New Jersey Institute of Technology, Newark, NJ 07102

Introduction

The presence of polychlorinated dibenzo-p-dioxins (PCDD) and dibenzofurans (PCDF) has been detected in fly ash from municipal solid waste incinerators (MSWI) as early as 1977 [1] and dioxin levels are consistently observed as a product from current MSW incineration, albeit at reduced levels relative to older combustors. There is, in addition, a recent enhancement in worldwide concern over emissions of PCDD and PCDF from combustion and industrial sources [2] and their levels are often monitored in some countries [3], with strict limitations (regulation) on effluent levels. The emission of dioxins from industrial sources are reported to be reduced by ca. 75% between 1987 and 1995 [4]. EPA's ability to understand risk and the EPA's evaluation of this risk over roughly the same time frame has also improved. A demonstration of this is a draft EPA report recently referenced by the media [5], which concludes (for the first time) that dioxin is a human carcinogen. This latest dioxin reassessment indicates that developing advanced methods of control of PCDD/F emissions remains important, even given the substantial reduction in industrial emissions.

PCDD/F emissions reduction over the past decade (~75%) can be attributed to recognition that *de novo* synthesis is the dominant mechanism of formation [6] and an understanding of the physical characteristics of this process, which is operative in the combustor's air pollution control devices. The PCDD/F formation is primarily associated with waste heat boilers and dust collectors in the air pollution control equipment of MSW incinerators[7-10]. Modifications of the air pollution control device operating parameters, such as temperature, has often resulted in significant emission reductions. Combustors with waste heat boilers have also had to install additional air pollution control equipment to reduce dioxin concentrations to regulatory levels.

This group of compounds, often symbolized by the terms "dioxins and furans" or PCDD/F's, consist of 75 isomers of polychlorinated dibenzo-p-dioxins and 135 isomers of polychlorinated dibenzofurans. The common units of their concentration measurement is somewhat unique, it is termed "Toxic

TABLE 4.1 Various Toxicity Equivalent factors (TEFs) for Specific PCDD/F Congeners [11].

Isomer	I/TEF[12]	EPA	BGA[13]
2,3,7,8 - TetraCDD	1	1	1
1,2,3,7,8 - PentaCDD	0.5	0.2	0.1
1,2,3,4,7,8 - HexaCDD	0.1	0.04	0.1
1,2,3,6,7,8 - HexaCDD	0.1	0.04	0.1
1,2,3,7,8,9 - HexaCDD	0.1	0.04	0.1
1,2,3,4,6,7,8 - HeptaCDD	0.01	0	0.01
OctaCDD	0.001	0	0.001
2,3,7,8 - TetraCDF	0.10.1	0.0.11	0.1
1,2,3,7,8 - PentaCDF	0.05	0.1	0.1
1,2,3,4,7,8 - PentaCDF	0.5	0.1	0.1
1,2,3,4,7,8 - HexaCDF	0.1	0.01	0.1
1,2,3,6,7,8 - HexaCDF	0.1	0.01	0.1
1,2,3,7,8,9 - HexaCDF	0.1	0.01	0.1
2,3,4,6,7,8 - HexaCDF	0.1	0.01	0.1
1,2,3,4,6,7,8 - HeptaCDF	0.01	0.001	0.01
1,2,3,4,7,8,9 - HeptaCDF	0.01	0.001	0.01
OctaCDF	0.001	0	0.001
Other PCDD/PCDF	0	0-0.01	0.001-0.01

Equivalent (TE)" and is calculated based on an assigned factor (TEF - Toxic Equivalency Factor) for the potential toxicity of each specific isomer in relation to 2,3,7,8 - tetrachloro-dibenzo-p-dioxin and the isomer's mass. The most common schemes for applying PCDD/F toxicity factors are shown in Table 4.1; the International Toxic Equivalent Factors "I/TEF" are the most widely used [11]. A Toxic Equivalent (TE) value of a mixture is calculated by multiplying the concentration (pg./g or ng/m^3) of individual congeners by their respective TEF. The sum of the TE concentrations for the individual congeners is the TE concentration for the mixture.

Reduction of PCDD/F compounds generated in the combustion process can be achieved by minimizing the formation of these products during the process and applying efficient removal and destruction technologies [14, 15]. An optimized (complete) burnout is one effective way to limit the formation of PCDD/F by reducing the unburned carbon species in the raw gas. Minimization of adsorption of incomplete combustion products on fly ash surfaces in the air pollution control equipment, where catalyzed formation of PCDD/F occurs is a second method for effectively limiting their formation. Control of dust release and prevention of large dust deposits in the boiler are also helpful to limit the PCDD/F formation.

PCDD/F removal in modern incinerators is performed by several techniques: adsorption on charcoal in a bed filter, adsorption on activated carbon injection into the flue gas, or by catalytic destruction in treatment of the flue gas [16]. Injecting of oxidizing agents into the raw gas between the boiler and filter is also utilized [17].

PCDD/F Concentrations in MSW Feed

The distribution of PCDD/F concentrations in MSW Feed was studied by categorizing the waste into four different municipal waste fractions [18]:

A = paper + cardboard;
B = plastics, wood, leather, textile etc;
C = fine debris < 8 mm;
D = food and garden wastes + screen remainder > 8 mm .

The results are shown in Table 4.2. The average PCDD/F concentration of dried (analytical) waste amounted to 104 pg. (TE)/g, or 50.2 (TE) pg./g of the wet total waste. A listing of actual waste sample data is summarized in Table 4.3 [18]. The average PCDD/F concentrations range from 2 to 50 pg. (TE-BGA)/g in data which is ten years old. Published data consistently indicate that PCDD/F concentrations in waste are about 50 pg.(TE)/g [19-21].

Lahl U. et al [20] collected and analyzed input samples of municipal waste from the MWI (Municipal Waste Incinerator) Bielefeld-Herford at 1989 - 1990. The results of sewage sludge and milled waste in the input which were collected at two sample collection episodes, one in summer and one in winter, are summarized in Table 4.4. The input showed a severe burden of PCDD/F in which a range of 10 to 41 pg (TE-BGA)/g PCDD/F was found. Several high concentrations (up to 7,655 ng/g) of PCDD/F precursors like PCB, chlorobenzenes and chlorophenols were also identified in the input (waste including sewage sludge).

TABLE 4.2 Range of PCDD/F Concentration in Different MSW Fractions [18]

Fraction	Waste Type	pg. (TE)/g
A	Paper, cardboard	18 - 383
B	Plastics, wood, leather, textile	29 - 1370
C	Fine debris (dust and particles) < 8 mm	8 - 468
D	Food and garden wastes, screen reminders > 8 mm	7 - 100

TABLE 4.3 PCDD/F Concentrations in Actual Waste Samples (1989 - 1991) [18]

Location	pg. TE (BGA)/g
Berlin, FRG* (1 sample) [22]	7.8
Berlin, FRG (green waste, 5 samples) [22]	2.1 - 13.6
Bielefeld, FRG (2 samples) [20]	33 - 41
FRG (average of 6 samples) [23]	50

* FRG = Former Federal Republic of Germany

TABLE 4.4 Results of two sample collection periods at the MWI Bielefeld-Herford [20]. S: Summer (27.7.1989), W: Winter (23.1.1990); ng/g referring to dry substance

PCDD/PCDF	sewage sludge ng/g S	sewage sludge ng/g W	milled waste ng/g S	milled waste ng/g W
Sum PCDD	5.15	2.552	13.21	7.939
Sum PCDF	< 0.01	0.757	0.3	1.397
PCDD+PCDF	5.15	3.309	15.51	8.79
TE (BGA)	0.01	0.021	0.03	0.041
PCB	139	47	514	527
Chlorobenzenes	86	18	154	128
Chlorophenols	3090	435	7655	1070

TABLE 4.5 PCDD/PCDF contents in different compost samples [20]

		PCDD µg/kg	PCDF µg/kg	PCDD + PCDF µg/kg	PCDD + PCDF ng (TE)/kg
A	Mixed waste compost	12.53	0.04	12.57	22.6
B	Mixed waste compost	19.1	0.04	19.14	32.1
C	Plant waste compost	1.62	< 0.01	1.62	1.8
D	Plant waste compost	2.94	0.16	3.1	5.2
E	Plant waste compost	1.88	< 0.01	1.88	2.2
F	Bark compost	2.15	0.05	2.15	3.6
G	vegetable waste compost	15.91	0.05	15.96	19.4
H1	vegetable waste compost	4.68 - 21.28	0.14 - 1.76	5.39 - 21.67	7.1 - 35.7
H2	vegetable waste compost	11.06	0.04	11.1	13.4
I1	vegetable waste compost	17.92	0.11	18.03	21.8
I2	vegetable waste compost	0.276	0.285	0.562	7.7
I3	vegetable waste compost	0.734	0.443	1.177	15.4
I4	vegetable waste compost	0.338	0.056	0.394	0.8
K1	Source not specified	8.543	0.884	9.427	30.8
K2	Source not specified	4.782	0.523	5.305	19.7
K3	Source not specified	5.394	0.534	5.928	24.2

Lahl U. et al [20] also analyzed several composts from the waste consisting of vegetable and other plant materials, which are usually about 30 to 40 percent of municipal solid waste. The result of PCDD/F contents in four different materials (green waste, plant waste, mixed waste, and bark) are shown in Table 4.5. The PCDD/F toxicity equivalents range from 0.8 to 35.7 pg (TE)/g.

PCDD/F Concentrations in Modern MSW Incinerator Effluent and Solid Streams

PCDD/F emission levels on the effluent of a MSW incinerator has been reduced remarkably over the past decade, due to improved combustion conditions, modern furnace designs, and adequate APC (air pollution control)

TABLE 4.6 Municipal Solid Waste Incinerator PCDD/F Emission Limits (in ng(TE)/m^3)

EU*	0.1
Germany	0.1
The Netherlands	0.1
USA	0.14 - 0.21
Japan	0.5

* EU = European Union

devices. The international air emission standards are summarized in Table 4.6. A modern MSW incinerator has no problem to reach the internationally accepted emission of 0.1 ng(TE)/m^3[14].

The PCDD/F concentrations in the different ash streams are also reduced from a modern MSWI. PCDD/F concentrations of 1 - 10 ng(TE)/kg are found in the bottom ashes [14]. This concentration level is as low as the PCDD/F level which are found in uncontaminated soils in Western Europe (5 ng(TE)/kg) [11]. The concentration of PCDD/F in boiler ashes and APC residues from a modern MSWI are also low. A typical boiler ash concentration 0.023 ng(TE)/g and APC residue concentration 0.213 ng(TE)/g are reported [11].

Do Modern MSW Incinerators Actually Destroy More PCDD/F Than They Produce

It appears that modern, efficient Municipal Solid Waste Incineration is typically a net-destroyer of PCDD/F. One estimates that an averaged annual quantity of 10 grams (TE) PCDD/F is in the feed (input) to a MSW waste incinerator that has an annual throughput of 2 * 10^8 kg. This 10 gram value is based on PCDD/F concentration of 50 pg.(TE)/g in table 4.3. The PCDD/F concentration ranges in the effluent and ash streams are determined based on the MSW output production rates, which are shown in Table 4.7. The accumulated annual output in both the gas and the solid mass streams (bottom ash and fly ash/air pollution control residue) are 0.44 - 1.67 g(TE), by assuming operation similar to that reported [24] for modern efficient MSW incinerators. This operation is designed for optimum combustion and for low emission of dioxins [14]. This data show that 16 - 23 times more dioxin (PCDD/F) is destroyed in the MSW than is produced in effluent to the environment. Again use of an efficient municipal solid waste incinerator is assumed.

The effective removal is actually more impressive than this. If (and this is a significant if) one assumes that the bottom ash, the boiler ash and the APC ash material are all removed from the accessible environment, that is they are stored in a well sealed and managed land fill, then only the effluent gases car any PCDD/Fs back into the environment. These effluent gases are reported in

TABLE 4.7 PCDD/F concentrations in input and output of a modern MSW Incinerator. (The calculation for MSW mass and PCDD/F mass are based on annual rate).

	Input	Bottom ash	Boiler ash	APC residue	Flue gas
MSW output production rate[11]		250 - 420 kg/Mg	2 - 12 kg/Mg	7 - 16 kg/Mg	4,000 - 4,500 m^3/Mg
MSW mass	2.0E08 kg	(5.0 - 8.4) E07 kg	(0.4 - 2.4) E06 kg	(1.4 - 3.2) E06 kg	(8.0 - 9.0) E08 m^3
PCDD/F concentration	0.05 ng(TE)/g	1 - 10 ng(TE)/kg	0.023 ng(TE)/g	0.213 ng(TE)/g	0.1 ng(TE)/m^3
PCDD/F mass	10 g(TE)	0.05 - 0.84 g(TE)	0.009-0.055 g(TE)	0.3 - 0.68 g(TE)	0.08 - 0.09 g(TE)

Table 4.7 to incorporate an average of 0.085 g(TE) PCDD/Fs. This is a reduction in over a factor of 100 from the input (10 g(TE) PCDD/Fs). The impressive destruction efficiency is therefore achieved using efficient incineration and modern, available air pollution control strategies to reduce PCDD/F formation as well as available and effective destruction technologies in the solid waste incineration effluent treatment.

Do Modern Waste Incinerators Actually Destroy More PCDD/F Than They Produce? The Answer Is – The Available Data Strongly Indicate This Is Correct; But More Study is Needed for further Verification.

References

Olie, P. Vermeulen and O. Hutzinger. Chlorodibenzo-p-Dioxins and lorodibenzofurans are Trace Components of Fly Ash and Flue Gas of e Municipal Incinerators in the Netherlands. Chemosphere 6: 455, 1977.

r, H., 1993, Organohalogen Compd., *11*, 221.

k and K. Jones, 1996, Dioxins in the Environment: A Review of ta, Environ. Sci. Technol. 30, 3133-3143.

1998, The Inventory of Sources of Dioxin in the United States; iew Draft; EPA/600/P-98/002Aa; The Office of Research and National Center for Environmental Assessment: Washington,

5. MSNBC News, May 17, 2000.

6. U.S. EPA Report, "Guidance on Collection of Emissions Data to Support Site-Specific Risk Assessments at Hazardous Waste Combustion Facilities," EPA530-D-98-002, 1998.

7. H. Tejima, I. Nakagawa, T. Shinoda, 1996, Chemosphere 32: 169-175.

8. T. Kato, S. Osada, S. Sakai, M. Hiraoka, 1996, Chemosphere 32: 145-150.

9. Y. Akimoto, S. Nito, Y. Inouye, 1997, Chemosphere 34:791-799.

10. T. Fujii, T. Murakawa, N. Maeda, 1994, Chemosphere 29:2067-2070.

11. International Ash Working Group: A. Chandler, T. Eighmy, H. Sloot, L. Velhow., 1997, Municipal Solid Waste Incinerator Residues, Elsevier, Amsterdam.

12. NATO CCMS, 1988, International Toxicity Equivalency Factors (I/TEF) Method of Risk Assessment for Complex Mixtures of Dioxins and Related Compounds, Report 178.

13. BGA, 1985, Sachstand Dioxine, Bericht des Umweltbundesamtes 5/85: 264.

14. J. Velhow and F. Karlsruhe. June, 2000, Low Cost, Low Pollution - Current Objectives in Waste Combustion, Proceedings of International Conference on Combustion, Incineration/Pyrolysis and Emission Control, Seoul, Korea.

15. D. Shin, J. Choi and S. Choi. June 2000. Interpretation of Design and Operation Parameters on PCDD/Fs Emission of a MSW Incinerator, Proceedings of International Conference on Combustion, Incineration/Pyrolysis and Emission Control, Seoul, Korea.

16. M. Hiraoka, 1989, Catalytic Destruction of PCDDs in Flue Gas, Chemosphere 19: 361.

17. H. Vogg and J. Velhow, 1991, Head-end-Techniken zur Dioxinminderung. VDI Berichte 895, 193.

18. M. Wilken, B. Cornelson, 1992, Distribution of PCDD/F and other Organochlorine Compounds in Different Municipal Solid Waste Fractions. Chemosphere 25: 1517-1523.

19. H. Tosine, R. Clement, V. Ozwacic and G. Wong, 1985, Levels of PCDD/F and other Chlorinated Organics in Municipal Refuse. Chemosphere 14: 821.

20. U. Lahl and J. Jager, 1991, PCDD/PCDF Balance of Different Municipal Waste Management Methods, Chemosphere 23: 1481.

21. B. Johnke and E. Stelzner, 1992, Results of the German Dioxin Measurement Programme at MSW Incinerators, Chemosphere 10:345-355.

22. ITU - Forschung measurements (Kolonnenstr. 26, D - 1000, Berlin 62, FRG).

23. H. Hagenmaier, unpublished (Institute of Organic Chemistry, University of Tubingen, D-7400 Tubingen, Federal Republic of Germany).

24. Environment Canada,1988, National Incinerator testing and Evaluation Program: Environmental Characterization of Mass Burning Incinerator Technology at Quebec City, Summary Report. Environment Canada Report EPS 3/UP/5.

ENERGY RECOVERY: AN ESSENTIAL TOOL OF INTEGRATED WASTE MANAGEMENT

Nickolas J. Themelis[1] and Young Hwan Kim[1,2]
[1]Department of Earth and Environmental Engineering, Columbia University;
[2]Department of Materials Science, Hong-Ik University, Seoul

ABSTRACT

The principal means for integrated management of municipal solid wastes (MSW) are recovery of materials (recycling), recovery of energy by combustion, bioconversion to fuel and compost, and landfilling of the remaining residues. This study examined the recovery of energy by using the combustible components of MSW as a fuel in a properly designed combustion reactor and thermoelectric plant to generate electricity and process steam. Despite the heterogeneity of materials in MSW, the mean hydrocarbon structure can be approximated by the organic compound $C_6H_{10}O_4$. A formula is derived that allows the prediction of the heating value of MSW as a function of moisture and glass/metal content and compares well with experimentally derived values. The performance of a leading Waste-to-Energy plant in the U.S. that processes about 0.9 million tons of MSW per year and produces a net 620 kWh/ton is examined. The results of this study indicate that energy recovery from MSW can reduce considerably the amount of land consigned annually to landfilling, reduce envireonmental impacts and decrease, to a small extent, dependence on fossil fuels.

Introduction

The municipal solid wastes (MSW) are the most non-homogeneous of all solids since they consisit of the residues of nearly all materials used by humanity: Food and other organic wastes, papers, plastics, fabrics, leather, metals, glass and miscellaneous other inorganic materials. Everything wears out gradually or abruptly and then ends up either in MSW or is discarded in land or water. The annual generation of MSW in the U.S. is about 0.7 metric tons (0.8 short tons) per capita.

Processing or disposal of MSW require what is called Integrated Waste Management (IWM): Separating the MSW into a number of streams each of which is then subjected to the most appropriate method of resource recovery. The separation of MSW components can take place at the source, i.e. households or businesses or at Materials Recovery Facilities (MRFs) where manual and electromechanical methods are used. There are four principal methods for resource recovery or disposal of MSW:

- **Recovery of materials:** Recovered paper, plastic, rubber, fiber, metal, and glass can be re-used to produce similar materials.
- **Recovery of energy:** Recoverable energy is stored in chemical form in all MSW materials that contain hydrocarbons; this includes everything except metals, glasses, and other inorganic materials (ceramics, plaster, etc.). By combusting such wastes, electricity and steam can be generated.
- **Recovery of soil and fuel:** The natural organic components of MSW (food and plant wastes, paper, etc.) can be composted aerobically (i.e., in the presence of oxygen) to

carbon dioxide, water, and a compost product that can be used as soil conditioner. On the other hand, anaerobic digestion or fermentation produces methane or alcohol and a compost product; this method provides an alternate route for recovering some of the chemical energy stored in the hydrocarbon fraction of MSW.

- **Landfilling:** Any fraction of the MSW that is not subjected to any of the above three methods, plus any residuals from these processes (e.g., ash from combustion) must be disposed in properly designed landfills.

The objective of this study was to examine the recovery of energy by sorting and pre-processing the combustible components of MSW and then using them as a fuel in a properly designed combustion vessel, similar to those used for generating electricity in fuel-fired power plants. This paper is part of a continuing joint study conducted by the Earth Engineering Center and the Center for Urban Research and Policy of Columbia University on alternatives for MSW management in New York City.

Disposal of MSW to landfills

Table 1 is based on data provided by the Council of Environmental Quality (1997) and shows that all four methods of managing MSW are used in the U.S. It is interesting to note that in the period of 1980-1996, the fractions of MSW recycled or combusted nearly doubled; also, the fraction of composted materials (consisting mostly of yard wastes) increased to 5.4% of the total MSW. However, landfilling remains the major means of disposition of wastes in the U.S. For example, New York City currently recycles about 20% of its MSW (0.6 million short tons) as paper, metal, glass and plastics; the remainder is landfilled at "tipping" fees that have tripled in the last few decades to the current fee of about $72/ ton for out-of-state disposal.

Table 1. U.S. Municipal Solid Waste Trends (Council for Environmental Quality, 1997)

	1980		1990		1996	
	10^6 tons*	%	10^6 tons*	%	10^6 tons*	%
Gross discards	151.64		205.21		209.66	
Recycling	14.52	9.6	29.38	14.3	46.01	21.9
Composting	<0.5	0.0	4.2	2.0	11.32	5.4
Combustion	13.7	9.0	31.9	15.5	36.09	17.2
Landfilling	123.42	81.4	139.73	68.1	116.24	55.4

*in short tons; for metric tons, multiply by 1.1.

Composition of MSW

The MSW composition varies amongst communities, and even within one community from year to year, but the differences is not substantial. The MSW composition used in this study is based on the New York citywide data of an extensive characterization study conducted in 1990 for the NYC Department of Sanitation (SCS Engineers, 1992). Table 2 shows that the major

Table 2. Composition of New York City MSW

WASTE COMPONENT	% WEIGHT*		tons/day **	
"Dry" Stream Combustibles		*51.9*		*6121*
Paper		31.3		3691
Corrugated Cardboard	4.7		554	
Newspapers	9.2		1085	
All other paper	17.4		2052	
Plastics		8.9		1049
HDPE (clear & color)	1.1		134	
Films and Bags	4.8		568	
PET	0.5		58	
Polypropylene, polystyrene	0.9		108	
PVC	0.1		14	
All other plastics	1.4		167	
Other dry combustibles		11.7		1380
Wood	2.2		259	
Textiles	4.7		554	
Rubber & Leather	0.2		24	
Fines	2.3		271	
Other	2.3		272	
"Wet" Stream Combustibles		***28.0***		***3302***
Food Waste		12.7		1498
Grass/Leaves		3.4		400
Brush/Prunings/Stumps		0.7		83
Disposable Diapers		3.4		401
Miscellaneous Organics		7.8		920
"Dry" Non-combustibles		***9.8***		***1156***
Glass		5.0		590
Clear Glass Containers		2.9		342
All other glass		2.1		248
Aluminum		0.9		106
Ferrous Metal		3.9		460
Other materials		***10.3***		***1215***
Hazardous Waste		0.4		47
Bulk Items (appliances, furniture, etc)		9.9		1168

* Adapted From: SCS Engineers, 1992 ** At current rate of generation of 11800 metric tons/day

constituent of MSW is paper. Other low-moisture combustible materials are plastics, textiles, rubber, leather, and wood. These materials can be called "dry combustibles", in distinction to the "wet combustibles" of food, plant and other wastes that contain 50-70% water. The third category shown in Table 2 is "dry non-combustibles", i.e. metal, glass and other inorganic compounds that have no heating value. Hazardous wastes, such as paints, oils, and chemicals constitute only 0.4% of the total waste stream and must be handled separately. Large items such as appliances and furniture can be broken down to metal scrap, or wood scraps that are either recyclable or combustible.

Separation of MSW into "dry" and "wet" materials

MSW can be classified into "dry" and "wet" materials, on the basis of their moisture content. The unpleasant odors and liquids associated with "garbage" are due to the putrescible organic components of food and plant wastes in the "wet" stream. These materials are less than 30% of the total MSW; yet they contaminate and complicate the transport and processing of the rest of the MSW. Therefore, it is generally preferable to separate the "wet and "dry" components at the source. This is already done at some forward-looking communities in Canada (Guelph, 2000; Halifax, 2000), Europe and Australia. There are three options for handling the "wet" fraction: combustion, aerobic or anaerobic bioconversion and landfilling. Table 3 (after Tchobanoglous et al, 1993) shows the "proximate analysis" composition of various types of combustible waste materials, i.e. % moisture, volatile matter (principally hydrocarbons), fixed carbon, and non-combustible, non-volatile "ash". The experimentally determined heats of combustion of each waste material are also shown.

Table 3: Proximate Analysis of Components of MSW (% weight)*

	Moisture	Volatile Matter	Fixed Carbon	Non-Combustible	kJ/kg As collected
"Dry" Combustibles					
Paper	10.2	75.9	8.4	5.4	15814
Cardboard	5.2	77.5	12.3	5.0	16380
Mixed Plastics	2.0	95.8	2.0	2.0	32800
Textiles	10.0	66.0	17.5	6.5	17445
Rubber	1.2	83.9	4.9	9.9	25330
Leather	10.0	68.5	12.5	9.0	18515
"Wet" Combustibles					
Wood	20.0	68.1	11.3	0.6	15445
Yard Wastes	60.0	30.0	9.5	0.5	6050
Food Wastes	70.0	21.4	3.6	5.0	5350
"Dry" and "Wet"	21	52	7	20	11630

* Adapted from data in Tchobanoglous et al., 1993.

Table 3 shows that wood has nearly the same heating value per unit mass as paper, while yard and food wastes contain less energy because of their high moisture content. For example, food wastes contain about 70% moisture and their calorific value is only 5350 kJ/kg (2300 BTU/lb). Thus, high-moisture food wastes contain enough heat to burn "autogenously" (i.e. without fuel addition) but cannot generate much electricity.

Thermal energy in MSW components

In this study, the chemical thermodynamics of the MSW combustion reaction were modeled by representing the composite combustible MSW by an established hydrocarbon compound. On the basis of the composition data for the "typical" U.S. MSW by Tchobanoglous 1993), and the atomic weights of the respective elements, Themelis and Kim (2000) calculated the molecular formula corresponding to each of the combustible components of MSW:

Mixed paper:	$C_6H_{9.6}O_{4.6}N_{0.036}S_{0.01}$
Mixed plastics:	$C_6H_{8.6}O_{1.7}$
Mixed food wastes:	$C_6H_{9.6}O_{3.5}N_{0.28}S_{0.2}$
Yard wastes:	$C_6H_{9.2}O_{3.8}N_{0.01}S_{0..04}$

They also showed that the hydrocarbon formula $C_6H_{10}O_4$ most closely approximated the mix of organic wastes in MSW. Representing the NYC dry stream by the $C_6H_{10}O_4$ formula results in the following equation for combustion:

$$C_6H_{10}O_4 + 6.5O_2 = 6CO_2 + 5H_2O \qquad (1)$$

This reaction is highly exothermic and at 1000°C generates about 27000 kJ/mol. Since the molecular weight of $C_6H_{10}O_4$ is 146, the "theoretical" heat of reaction (i.e. in the absence of inert or moisture) per unit mass of MSW is calculated to be 18400 kJ/kg (7900 BTU/lb). Similarly, if the MSW combustibles are simulated by the less oxidized compound $C_6H_{10}O_3$, the combustion reaction is

$$C_6H_{10}O_3 + 7O_2 = 6CO_2 + 5H_2O \qquad (2)$$

and the "theoretical" heat of combustion is 23000 kJ/kg (9900 BTU/lb.).

Effect of moisture and inert materials on available heating value

The inclusion of moisture and non-combustible materials in the MSW affects the available heat in Waste-To-Energy (WTE) plants that produce electricity and steam. To quantify these effects, it is assumed that the WTE plant provides steam to a standard power plant and that the exhaust gases leave the boiler at 120°C and 135 kPa (20psi). Accordingly, the amount of heat wasted per kg of water in the feed, as water vapor in the exhaust gases, is calculated to be 2636 kJ/kg. The non-combustible materials in the feed, mainly glass and metals, will end up mostly in the bottom ash. If it is assumed that the ash leaves the grate at about 700°C and a reasonable value for the specific heat of ash, the corresponding heat loss to inorganic materials fed with the combustibles is estimated to be as follows:

- Glass and other siliceous materials: 628 kJ/kg (270 BTU/lb.)
- Iron: 420 kJ/kg
- Aluminum: 1134 kJ/kg

Since the iron/aluminum ratio in MSW is about 4 (Table 3), the mean heat loss per kg of metal is estimated to be 544 kJ/kg (234 BTU/lb.). Accordingly, the effects of non-combustibles on the heating value of RDF can be expressed as follows:

Heating value of mixed MSW = (heating value of combustibles)*X_{comb} –
- (heat loss due to water in feed)*X_{H2O} – (heat loss due to glass in feed)* X_{glass} –
- (heat loss due to metal in feed)*X_{metal} (3)

and substituting numerical values for heating value of Equation 1 and the above heat loss gives,

Heating value of mixed MSW

$$= 18400X_{comb} - 2636X_{H2O} - 628X_{glass} - 544X_{metal} \quad kJ/kg \quad (4)$$

where X_{comb}, X_{H2O}, etc. are the fractions of combustible matter, water, etc. in the RDF.

From the point of view of combustion efficiency and maximization of energy recovery per ton of MSW, it is clear that it would be preferable to separate wet putrescible materials. This will increase the heating value of the material being burned and therefore will generate more energy per ton of waste burned. Separation of non-combustibles in the waste (glass and metals) will also increase heating value and energy generation per ton. However, the prime function of waste combustion facilities is waste disposal. The removal of specific materials before combustion requires either collection of separate streams at the source, or separation at the combustion facility before burning. Either of these processes increases the collection, transportation and processing costs with relatively little change in the energy generation per ton of original waste. Thus, although heating value is increased, the net cost of energy production from solid wastes also increases.

Table 4 shows that the experimentally determined heating value of the "dry" stream, after separation of the "wet" and the non-combustible fractions, amounts to 18470 kJ/kg. This value is fairly close to the thermochemical value calculated on the basis of Equation 1. Also, this value is in the range of lignitic and sub-bituminous coals that are still used in many power plants. Therefore, using the "dry combustible" MSW as a fuel can reduce, by as much as ton per ton, the need for mining coal. Another way of expressing the value of MSW as an energy resource is by stating that the $1.15x10^5$ GJ/day of heat contained in the NYC dry combustibles corresponds to about 3.05 million liters (806400 gallons) of No.2 fuel oil per day.

It is also of interest to compare the experimentally determined heating values of Table 4 with the projection of Equation (4) for the effects of moisture and inert on the heating value of RDF. Table 4 shows that when the dry non-combustibles are included, the combined stream consists of 84.1% combustibles, 7.8% metal and 8.1% glass and the experimental heating value is 15817 kJ/kg (6800 BTU/lb). This value is in close agreement with the value of 15400 kJ/kg (6620 BTU/lb.) projected by Equation 4. Also, if we assume that the "wet" fraction contains 50% water, when the wet fraction is commingled with the dry fraction, the composition of the entire stream becomes: Combustibles: 67.5%; glass and metal: 10.9%; moisture: 15.6%. Table 4 shows that the experimentally determined heating value of this mixed stream is 12332 kJ/kg, in comparison to the 12000 kJ/kg projected by Equation 4.

Table 4. Heating Value of Various Components of NYC MSW

Component of MSW Dry Stream	% in NYC Waste[1]	Weight[2] (Tons/day)	Heating Value[3] (MJ/ton)	Heating Value of Component (GJ/day)
Combustibles:				
Paper:	31.3			
Newspaper	9.2	1087	18548	20162
Cardboard	4.7	555	16382	9092
Other Paper	17.4	2056	15814	32514
Plastics (HDPE: 1.1%; PET: 0.5%)	8.9	1052	32800	34506
Textiles	4.7	555	17445	9682

Rubber & Leather	0.2	24	21387	513
Wood	2.2	260	15445	4016
Fines	2.3	272	8534	2321
Other	2.3	273	6978	1905
Total Dry combustibles	**51.9**	**6134**	**18470**	**114711**
Non-combustibles				
Metal	4.8	567		
Glass	5.0	591		
Total Non-combustibles	**9.8**	**1158**		
Total Dry Stream	*61.7*	*7292*	*15817*	*115338*
Total Wet Stream				
Including Food & yard waste, All other "wet"	*28.0*	*3309*	*4652*	*15393*
Total Wet+Dry[4]	***89.7***	***10601***	***12332***	***130731***

[1]SCS Engineers, 1990

[2]Weight is based on current NYC generation levels of 13000 short tons per day (stpd)

[3][illegible]

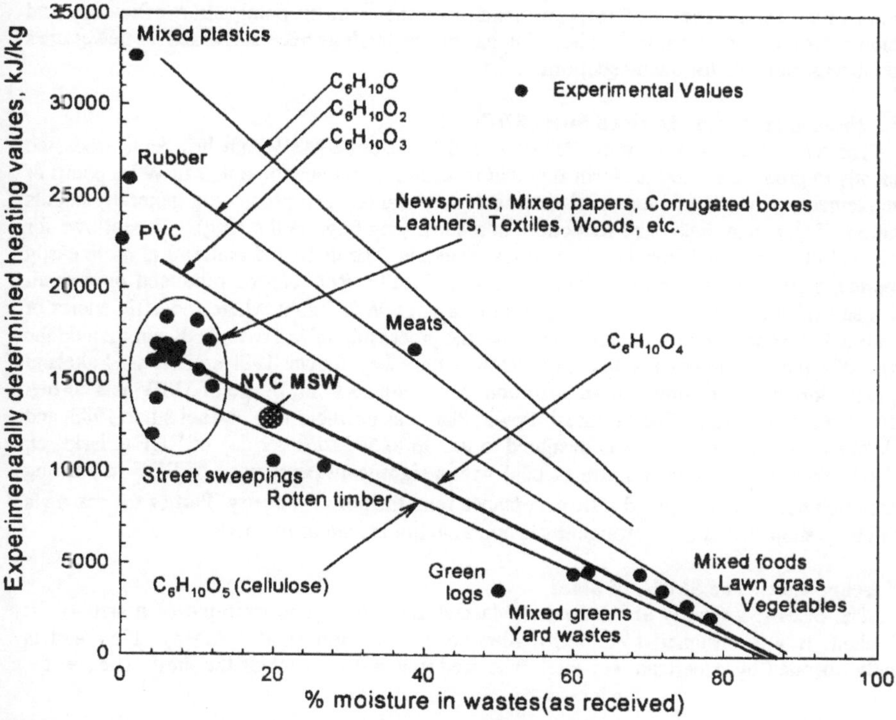

Figure1. Comparison of experimental heating values of various waste

WTE technologies: Mass Burn WTE plants

When people discuss the pros and cons of "incineration" of MSW, they are usually influenced by the results of past practice when there was no or little concern about the environment and emission controls were rudimentary or non-existent. For instance, between 1910 and 1968 there were approximately 17000 apartment/house incinerators operating in New York City; in the same period 32 municipal incinerators were constructed in NYC and combusted a total of about 73 million tons (81 million short tons) of MSW (Walsh et al, 2000).

More recently, "mass burning" incinerators were provided with emission control systems but many were not used for energy recovery. Today, mass burning of MSW with power generation, in Waste to Energy (WTE) plants is practiced widely in the U.S. and other developed countries and does not requires pre-processing of the waste, apart from manual removal of bulky items like "white goods". The rest of the waste is charged to the furnace by means of a mechanical "claw" that deposits full garbage bag and other items at the feed end of a metal grate that moves the waste materials slowly through the combustion chamber. Because of the large size of the items deposited on the grate mechanism, the rates of mass transfer and chemical reaction with the oxidizing gas are relatively slow. Because of the low rate of oxidation, a very large combustion chamber and grate are required and the intensity of combustion (rate of heat generation per unit volume) is correspondingly low. Therefore, the temperatures generated in the combustion chamber are in the order of 900°C and the ash does not reach the point of fusion or semi-fusion. There have been improvements over the years in combustion efficiency and pollution control of mass burn WTE plants but has this technology been surpassed by newer ones that are discussed in the following sections.

WTE technologies: Refuse Derived Fuel (RDF)

The term refuse-derived fuel (RDF) is used to describe MSW that has been processed sufficiently to produce a fairly uniform fuel that is ready for combustion either in WTE plants or as complementary fuel in conventional thermoelectric plants. The processing generally entails separation of inert materials, size reduction, and densifying (e.g., pelletizing). This allows for the removal of both recyclables and hazardous materials. The densified material is more easily transported, stored, and combusted than raw MSW. Ideally, RDF can be produced on a small scale at several locations and then transported to a large WTE plant where the efficiencies of scale allow for effective emission controls. Also, the processing of MSW to RDF can include the addition of calcium compounds that after combustion they reduce HCl emissions. Lakeland Electric in Florida is an example of an operation that continues to make use of MSW as a co-fuel in a coal-fired power plant. The McIntosh Power Plant has been burning co-fuel since 1983, and uses 10% RDF to 90% coal. It was designed to use up to 500 tons per day of RDF (Clarke, et. al., 1991). However, the combination of coal with a significant percentage of RDF in existing power station boilers has received little acceptance from the power industry. Part of the reason is that WTEs is subjected to more stringent gas emission limits than utility boilers.

WTE technologies: The SEMASS plant

The SEMASS facility at Rochester, Massachusetts is a good example of a state-of-art WTE plant. It was designed by Energy Answers Corporation (EAC, Albany, NY) and is presently operated by American Ref-Fuel. The feed material consists of the entire (i.e., wet +

dry) MSW stream. The facility takes in waste by covered rail car and truck from 50 communities in a 65-mile radius, including the entire Cape Cod area and Martha's Vineyard. The plant consists of three parallel combustion units and processes over one million short tons per year. The first two units were built in 1989 and Unit 3 in 1994.

Waste brought to plant is dumped on a tipping floor (Figure 2). The waste is loaded onto conveyors that pass by inspectors who look for bulk waste that could jam the shredders, or for hazardous waste; these items constitute about 1.6% of the incoming material. The waste is then shredded in one of two large hammermill shredders that produce a blended material of –6 inch size coarse RDF. The shredded material is conveyed under overhead belt magnets for the first round of ferrous metal recovery and is then stored in bays in a closed building. This material is called "processed refuse fuel" (PRF) and apparently can be stored in a storage building for long periods. Because some of the moisture has evaporated during shredding and storage, any moisture left is distributed and absorbed throughout the PRF; therefore the PRF does not have the acrid smell of garbage and does not attract rats or flies.

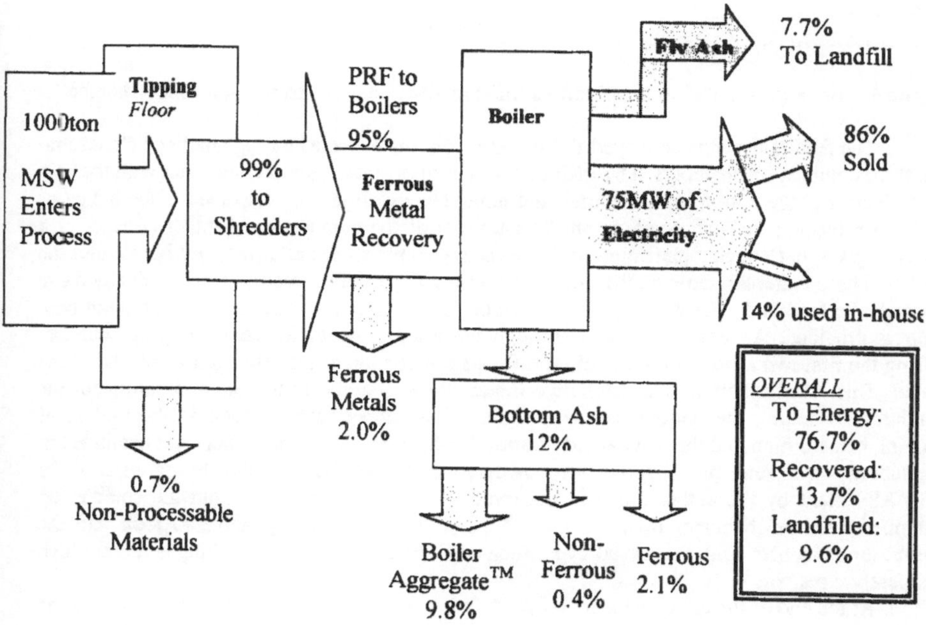

Figure 2. Material flows in and out of SEMASS combustion chamber

Combustion chamber Figure 3 shows a computer simulation of gas velocity profiles in the SEMASS combustion chamber. The inside dimensions of the chamber are 6m deep (grate length) by 11 m wide by 30 m high.

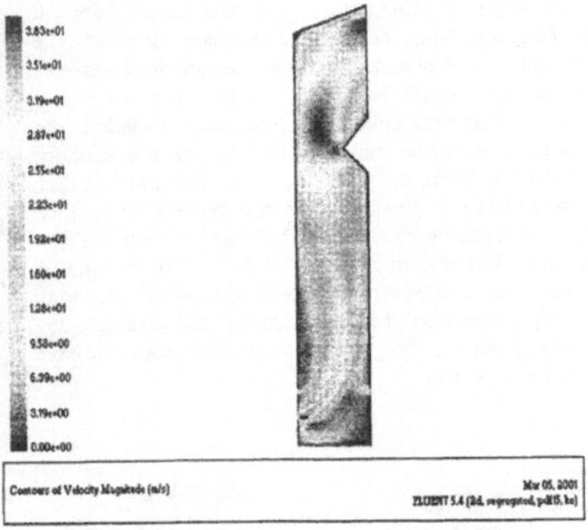

Figure 4. Computer simulation of velocities (m/s) in 30-m high SEMASS combustion chamber

The feed material is conveyed to bins and from there ejected through inclined chutes into the three combustion chambers. The PRF contains a lot of small, light particles that with the help of deflectors at the bottom of the chutes and high-velocity air jets are dispersed in the hot gases and are subjected to flash combustion. The bottom of the combustion chamber consists of a moving grate that collects inert materials and heavy combustibles after they are blown into the boiler. These materials settle on the end of the grate away from the feed end and gradually move towards the feed end. The grate speed is variable but generally a residence time of about one-hour is provided. An upward airflow through the grate provides for completing combustion, during the first two thirds of the travel, and for partial cooling of the ash towards the end of the travel. This counter-current heat exchange mode allows good heat recovery and eliminates the need for quenching the bottom ash with water. The temperatures reached in the middle of moving bed are high and the discharged "bottom" ash, collected by the authors and examined in the laboratory, is semi-fused, unlike the powdery ash of mass-burn plants. During visits of the SEMASS plant by the authors, it was observed that the temperatures within the combustion chamber were much higher than for mass-burn plants. It was not possible to look into the combustion chamber with the naked eye, much as is the case in copper smelting and other high temperature reactors.

At the end of the grate, the ash "clinker" falls onto a conveyor belt that transports it to the ash processing section of the plant. Ferrous metals that were not recovered before combustion and also non-ferrous metals are recovered from this ash by means of magnetic and eddy current separators, respectively. EAC has used the resulting granular aggregate as a substitute for crushed stone in concrete and asphalt applications and is presently using it as a government-mandated daily landfill cover.

The fly ash collected in the fiber bag filters contains most of the heavy metals that were present in the waste and is landfilled. EAC uses a patented stabilization system to bind heavy metals so that they do not leach out during various tests performed on the ash.

The SEMASS boilers are of the water-wall type without refractory linings. The superheated steam is used for the generation of electricity. The three units of the SEMASS plant process about 910000 metric tons per year and produce up to 720 kWh of electricity per ton of MSW, of which about 100 kWh/ton are used to run the plant and the remainder are sold to the local utility.

The gas handling plant of the SEMASS Combustion Unit 3 (built in 1994) is more advanced than the earlier two units. Potential air contaminants are controlled by a variety of means. A solution of urea in water is injected continuously into the furnace to control the level of nitrogen oxides. After the combustion gases leave the boiler they pass through water and air heat recuperators. Then they enter a "dry scrubber" chamber where a lime slurry is injected to neutralize acid gases and trap any chlorides and dioxins/furans that may have either persisted the high temperature atmosphere in the combustion chamber, or re-formed during the cooling stage of the gas. Finally, fiber fabric filter captures fine particles before the gases are discharged through the stack.

On the basis of reported operating data (Figure 2), the EAC system at the SEMASS facility, converts approximately 76.7% of the incoming waste into energy, recovers 4.5% as ferrous and non-ferrous metal, and disposes 7.7% as fly ash to the nearby backup landfill. The bottom ash, after metal recovery, represents about 10% of the feed. The combustion and power generations are continuous except for planned maintenance shutdowns. The fuel preparation stage (shredding and magnetic separation) operates only two shifts and the ash processing one shift each day.

WTE Gas Emissions

The most contentious issue in discussions of energy recovery from solid wastes is that of emissions to the atmosphere. Emissions of mercury, hydrochloric acid, and dioxins have been the most worrisome problems in the past. However, in modern WTEs such as the SEMASS No. 3 plant, they have been reduced to very low levels by means of reduction of the precursors in the feed, better combustion practices, and much improved gas control systems that include advanced dry-scrubbing and filter bag technologies. In particular, the problem of mercury emissions in the U.S. has subsided as this metal has been replaced by other metals in household batteries and by digital technology in thermometers, and by separate collection of fluorescent lights.

Table 5 compares air emission levels for the SEMASS No. 3 plant in Massachusetts with the current EPA standards. It can be seen that the emission levels actually attained in modern WTE facilities are substantially lower than those expected by the EPA. The volume of process gas in the SEMASS reactor was calculated to amount to about 6700 standard cubic meters per ton of MSW combusted. On this basis, the total (i.e., three combustion chambers) annual emissions of SEMASS amount to only 34 kilograms of mercury and 5 grams of dioxins. For comparison the total mercury emissions from all U.S, power plants in 2000 were about 42,700 kilograms (EWG 2001). As noted earlier, in the SEMASS plant, the high temperature combustion results in destruction of any dioxins/furans in the PRF; also dry scrubbing and then collecting fine particles by means of fabric filters combats the formation of these compounds after the boiler. It is also interest to note that the undesirable chloride ions enter the waste stream principally in the form of chlorinated organic compounds like PVC. Yet this material constitutes only 0.1% of the total waste stream and could easily be replaced by other materials in all current applications, as was done for the halide compounds that threatened the ozone layer, if a higher priority is assigned by EPA to recovering energy from MSW.

Table 7: Emissions from WTE Facilities Compared to EPA Standards

Emission	EPA Standard[1]	SEMASS[2]
Particulate (gr/dscf)	0.010	0.002
Sulfur Dioxide*	30	16.06
Hydrogen Chloride*	25	3.6
Nitrogen Oxides*	150	141
Carbon Monoxide*	150	56.3
Cadmium**	20	1.24
Lead**	200	30.03
Mercury**	80	5.09
Dioxins/Furans (ng/dscm)	30	0.86

gr/dscf: grain/dry standard cubic foot; 1 gr/dscf = 2.28 g/dscm
* ppmdv: parts per million dry volume
** μg /dscm: microgram per dry standard cubic meter; ng: nanogram
The standards and data are reported for 7% O_2, dry basis, and standard conditions.
[1]40 CFR Part 60, Subpart EB for new RDF-fired Municipal Waste Combustors.
[2]EAC, average of 1994-1998. Boiler No.3
[3]Studley, 1997.

Use or Disposal of Ash

Companies and researchers have been investigating ways of treating ash residues from WTE facilities. Ash consists of residues left in the combustion chamber (bottom ash) and in the air pollution treatment devices (fly ash). The post-treatment of ash produced by the low temperature combustion chambers such as fluidized beds usually involves vitrification at high temperatures in order to immobilize the metals. The main aim of ash treatment is to prevent the toxic constituents of the ash, especially dioxins, furans and heavy metals, from escaping into the environment after disposal. Solidification by means of vitrification or the application of various chemicals is further means of decreasing the chances of leaching metals. Phosphate has been shown to stabilize heavy metals in dusts that result from the vitrification of incinerator ash (Eighmy et. al., 1998). Treatment of ash is a much more mature technology than re-use.

The bottom ash produced in the SEMASS plant resembles clinker ash and, after mechanical separation of ferrous and non-ferrous metals, has a relatively high specific gravity (typically 2.25) and as per SEMASS reports contains less than 2% carbon and less than 1% fines. The toxicity characteristics leaching test (TCLP, Wiles, 1997) based on the EPA standard has shown that the metals in bottom ash are not leachable.

Comprehensive assessments of the re-use of ash from WTE plants are presented in the papers by Wiles (Wiles, 1995) and Chang (1998).

Life cycle assessment of energy recovery

A critical analysis and comparison of the life-cycle environmental impacts of recycling, incineration and landfilling by Dennison (1996), based on the major North American studies of the subject showed that recycling of used materials is superior to either incineration or landfilling, as one might reasonably expect. However, it also showed that combustion is preferable to landfilling and offered the advantage of 28 MJ less of energy used per ton of material processed than landfilling. Of course, interstate transport of MSW, as planned by NYC, will increase considerably the energy usage per ton of MSW.

Of equal importance was the finding that, for the 10 major air pollutants categories considered, combustion resulted in lower emissions than landfilling, with the exception of higher generation of carbon dioxide in the combustion process. However, when the generation of carbon dioxide is associated with the production of useful energy, e.g. 620 kWh/ton in the SEMASS process, one should subtract the "avoided" amount of carbon dioxide that would be produced anyway in a conventional power plant.

A more recent study by Eschenroeder (1998) indicated that, for modern landfills equipped with methane collection systems (70-year post-closure period), the time-integrated effect of green house gases emitted is 45 times greater than when the MSW is combusted.

Conclusions

This paper is part of a continuing joint study between Columbia University's schools of engineering and Public Affairs that entails an examination of both technology and policy constraints. The joint study recognizes that waste minimization by means of better design of products and packaging is highly desirable. Also the best means way of managing municipal solid wastes is by recovering recyclable materials. However, even when the best available technology is applied for collecting, sorting and processing, there remains a large fraction of wastes that cannot be recycled. It is then advisable, both from the economic and environmental viewpoints, to recover energy from the combustible fraction of MSW instead of consigning it to landfills.

It was shown that the formula $C_6H_{10}O_4$ is fairly representative of the organic mix in a typical MSW and that the calculated thermochemical heat of combustion of this compound is in good agreement with experimentally determined values for mixed organic materials. On this basis, a formula was derived that projects the heating value of MSW as a function of moisture and glass/metal content. The calculated and experimental heating values of the "dry combustible" fraction of MSW is in the range of lignitic and sub-bituminous coals that are still used in many power plants. Thus, using the "dry combustible" MSW as a fuel can reduce, by as much as ton per ton, the need for mining fossil fuels. For example, the usable energy in the New York City MSW is equivalent to nearly three million liters of fuel oil per day.

REFERENCES

Brady, M.H. (2000) Materials and Energy Recovery from the Dry Stream of New York City's Residential Waste, *M.S. Thesis*, Earth and Environmental Engineering, Columbia University, NewYork, NY 10027.

Chang, Ni-Bin, et al. (1999) The Assessment of Reuse Potential for Municipal Solid Waste and Refuse- Derived Fuel Incineration Ashes. Resources, Conservation and Recycling. 25

, 255-270.
Clarke, Marjorie J., Maarten de Kadt, David Saphire. (1991) Burning Garbage in the US: Practice vs. State of the Art. INFORM, Inc.: New York.
Council on Environmental Quality (1997) available on the White House web site (http://www.whitehouse.gov/CEQ).
Dennison, R.A. (1996) Environmental life-cycle comparisons of recycling, landfilling, and incineration, *Annual Revues of Energy and Environment*, Vol. 21, 191-237.
Department of Sanitation, NYC. (1998) 2001 and Beyond: A Proposed Plan for Replacing Fresh Kills Landfill. Michael T. Carpinello, Acting Commissioner.
Eighmy, T. Taylor, et al. (1998) Characterization and Phosphate Stabilization of Dusts from the Vitrification of MSW Combustion Residues. Waste Management. 18 : 513-524.
Energy Answers Corporation (EAC). (1999) Albany, NY (SEMASS Plant in Rochester, MA).
Eschenroeder, A. (1998) Greenhouse Gas Dynamics of Municipal Solid Waste Alternatives, presented at Air & Waste Management Assoc. conference on GLOBAL CLIMATE CHANGE, Oct. 13-16
EWG 2001 Environmental Working Group, www.ewg.org
Franklin Associates (1995) The role of recycling in integrated waste management in the United States, Rep. *EPA/530-R-96-001*, USEPA, Munic. Industrial Waste Division, Wash. DC.
Harrison, Brenda, P. Aarne Vesilind. (1980) Design & Management for Resource Recovery: Volume 2: High Technology-A Failure Analysis. Ann Arbor Science Publishers, Inc.: Ann Arbor, MI.
Horner & Shifrin, Inc. (1970) Study of Refuse as a Supplementary Fuel for Power Plants. St. Louis, MO.
Ohlsson, O. (July 1994) Results of Combustion and Emissions Testing When Co-Firing Blends of Binder-Enhanced Densified Refuse-Derived Fuel (b-dRDF) Pellets and Coal in a 440 Mw Cyclone Fired Combustor: Volume 1, Test Methodology and Results. Joint Effort of Argonne National Laboratory and National Renewable Energy Lab. NICH Report No. TP-430-6322a.
Roinen, A. (1999) *HSC Chemistry 4.0*, Outokumpu Research, Outokumpu Oy, Pori, Finland.
SCS Engineers. (1992) New York City Waste Composition Study,1989-1990
Serumgard, John. (1997) Scrap Tire Derived Fuel: Markets and Issues. Presentation at Fifth Annual North American Waste to Energy Conference. April 1997.
Studley, Bruce C. (1997) Start-up and Initial Operating Experiences at the Robbins Resource Recovery Facility while Maximizing Environmental Performance. Presented at Fifth Annual North American Waste-to-Energy Conference. April 1997.
Sutin, Gordon L. (1999) Various Company Reports. , Energy Answers Corporation, Albany, NY.
Tchobanoglous, G., Theisen, H., and Vigil, S. (1993) *Integrated Solid Waste Management*, Chapter 4, McGraw-Hill, New York.
Themelis, N.J., Y.H. Kim (2000) Materials and Energy balances in a Large –Scale Aerobic Bioconversion Cell. submitted for publication to Journal of Waste Management and Research, Sept.2000
U.S.EPA report,(1997) EPA 530-S-97-015 Characterization of Municipal Solid Waste in the United State 1996
Walsh, D.C. (1989) Solid Wastes in New York city; A History, Waste Age, 20(4), 114-124
Walsh, D.C., Chillrud, S.N., Simpson, H. J., and Bopp, R.F. "Refuse Incinerator Particulate

Emissions and Combustion Residues for New York City during the 20th Century under review Environmental Science and Technology (internal report available from Earth Engineering Center, Columbia University.

Wiles, Carlton C. (1996) "Municipal Solid Waste Combustion Ash: State-of-the-knowledge." Journal of Hazardous Materials. 47 , 325-344.

Wiles, Carlton C. (1997)"Questions and Answers about Waste-to-Energy Ash" The National Renewable Energy Laboratory, NREL/BR-430-22712

INVESTIGATION OF TIRE SHREDS FOR USE IN RESIDENTIAL SUBSURFACE LEACHING FIELD SYSTEMS: A FIELD SCALE STUDY

SUKALYAN SENGUPTA
Civil & Environmental Engineering Department
University of Massachusetts Dartmouth
North Dartmouth, MA 02747

HEATHER MILLER
Civil & Environmental Engineering Department
University of Massachusetts Dartmouth
North Dartmouth, MA 02747

INTRODUCTION

Scrap tires pose a significant environmental and public health problem. A number of applications of this material have been attempted, such as tire-derived fuel, artificial reefs and breakwaters, leachate collection systems in landfills, highway applications, etc. but they cannot recycle/reuse the enormous amount of tires discarded every year, much less the stockpiles that exist from past disposal. An interesting application area could be the use of scrap tire chips in lieu of crushed stone aggregate in residential subsurface leaching field systems. In terms of regulatory approval, the two primary concerns are i) does the material leach contaminants from its surface which can affect groundwater quality, and ii) does the material provide similar performance as a conventional leaching field system. This study details the results of field-scale experiments conducted to answer these questions.

BACKGROUND

It is estimated that about 250 million tires are discarded annually in the United States. Tires are manufactured from rubber that is sulfur-vulcanized. This process of vulcanization changes the rubber from a thermoplastic, which can be reprocessed many times, to a thermoset, which can be shaped only once [1]. This is the primary reason why scrap tires cannot be reused/recycled easily. Scrap tires represent a significant municipal solid waste problem because they are often placed in stockpiles which pose fire hazards and serve as breeding grounds for rodents and mosquitoes. In lieu of stockpiling scrap tires or placing them in already overcrowded landfills, a number of applications that reuse/recycle them have been attempted. These include tire-derived fuel, recycled rubber products, artificial reefs and breakwaters, daily cover and/or leachate collection systems in landfills, and numerous highway applications (crumb rubber asphalt pavement, lightweight fill for embankments and behind bridge abutments, roadbed insulation, etc.). A potential end market, which has not yet received much attention in most states, is residential subsurface leaching field systems. In the Commonwealth of Massachusetts, it is estimated that six million tires are discarded annually [1]. If Massachusetts codes were to allow the use of tire chips in residential septic systems, the potential would exist for about 25% of its scrap tires to be recycled and reused in these systems (assuming about 2000 tire chip systems were installed annually).

Although several states now have guidelines regarding the use of tire chips in residential subsurface leaching field systems, very little research has actually been completed to study the performance of these systems, and to investigate the potential for leaching of contaminants from the tire shreds in these systems. Therefore, the purpose of this project was to conduct an investigation of the use of tire chips instead of gravel in residential subsurface leaching field systems.

LITERATURE REVIEW

Humphrey [2] summarized hydraulic conductivity (permeability) values for tire shreds and concluded that it is generally much greater than most granular soils. Hydraulic conductivity is not an inherent property, but will vary depending upon numerous factors, including particle size distribution, porosity, and applied normal load (especially for tire shreds, which are very compressible). The various researchers tabulated in Humphrey [2] report hydraulic conductivity values ranging between 0.58 cm/sec and 23.5 cm/sec for tire shreds between 0.75 and 3 inches in size, indicating that tire shreds should provide more than enough flow capacity for use in septic systems.

To address the second, and the more critical, issue of potential groundwater contamination from leachates, there have been some studies published in the literature but none was conducted on septic tank effluent. The most significant of these include references [3-7]. A detailed review of these (and other) studies is presented in [1] and will not be repeated here, but a brief summary follows. In general, these studies indicated that leachate derived from scrap tires is "safe," i.e., most pollutants leached from tire shreds were generally below the various regulatory limits, except in studies conducted under extreme pH conditions. References 3 & 4 indicate that, in the laboratory, metals are leached at higher concentration under low pH conditions, while organics are leached at higher concentration under more basic conditions. Although none of the leachate samples exceeded the Extraction Procedure (EP) toxicity criteria or the TCLP criteria, the following metals exceeded the Minnesota Recommended Allowable Limits (RALs) for drinking water: arsenic, cadmium, chromium, selenium, and zinc. The iron level was also much higher than the Secondary Maximum Contaminant Level (SMCL). It should be noted that these studies were conducted under extreme pH conditions to evaluate worst-case scenarios and may not be representative of actual conditions in "scrap tire" application areas.

FIELD SCALE STUDIES

A field-scale study was conducted at the Massachusetts Alternative Septic System Test Center, located in the premises of the Massachusetts Military Reservation Base in Falmouth. Three leaching field trenches were installed; one of them being a conventional (gravel-filled) Title V system which conformed to Massachusetts Department of Environmental Protection regulations [8], and the other two containing tire shreds.

A schematic, showing the major components of each test pit and approximate dimensions is shown in *Figure 1*.

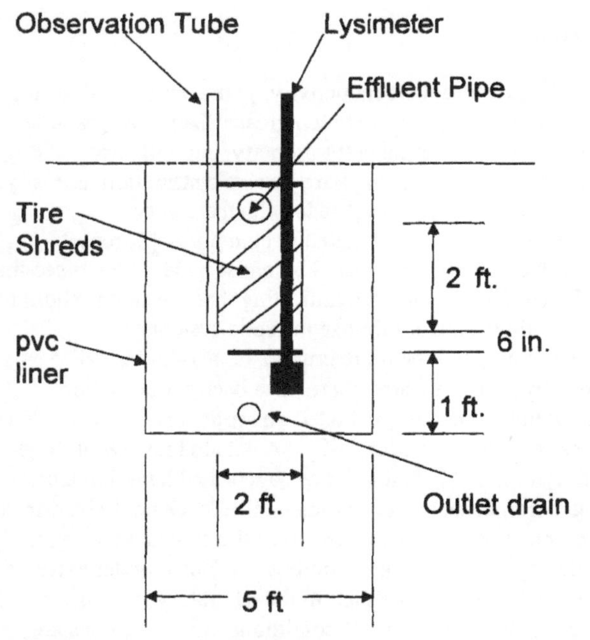

Figure 1: Typical Cross-Section of Leaching Field Trench (not to scale)

The initial construction involved excavating three large pits (one for each of the two tire-shred systems, and a third for the conventional gravel system). Each pit was about 5 feet wide by 15 feet long and about 4 feet deep. Each pit was lined with a 30 mil PVC geomembrane so that all effluent could be collected and removed through an outlet drain and then sent to the on-site wastewater treatment plant. Connected to the outlet drain was an underdrain system consisting of a 4-inch diameter perforated PVC pipe, surrounded by a layer of drainage stone. Each pit was designed to accommodate a bed of test material (tire shreds or gravel) 10 feet long by 2 feet wide by two feet high. To maintain consistent soil conditions surrounding the test material, each pit was filled with sand that was placed and compacted beneath and around the sides of the test material. Pan Lysimeters were fabricated and installed to collect effluent after it passed through the test material (tire shreds or gravel). The lysimeters consisted of a length of 2-inch diameter PVC pipe, with a sump attached at the bottom. A rectangular sheet of 40-mil PVC geomembrane was attached to the pipe with a collar located above the sump. Several small holes were drilled in the PVC pipe just above the collar. One lysimeter was installed in each test trench. The lysimeters were installed by embedding the sump in the sand layer, so that the PVC "pan" was located about a foot above the bottom of the pit. The sand beneath the pan was graded so that the pan sloped down towards the center, to allow effluent that collected in the pan to drain through the holes in the PVC pipe and into the sump. Coarse aggregate (drainage stone) was placed around the holes in the PVC pipe for proper filtration (i.e., to prevent the smaller sand particles from passing through the holes in the pipe and into the sump).

After the lysimeter was in place, a layer of sand, about 6 inches thick, was placed and compacted above the level of the pan. The base of the test trench was located on the top of this sand surface in each pit. Since construction of the test trenches was from the bottom up, it was necessary to construct a wooden form into which the test materials (tire shreds or gravel) could be placed, so that the sides of the trenches would be vertical. Two observation tubes were installed in each trench, each consisting of sections of 4-inch diameter PVC pipe, open at each end, with a removable cap at the top. The purpose of these tubes was to provide a means of observing the sand/tire shred or sand/gravel interface in each trench, check for potential build-up of deleterious material (such as iron deposits, etc.) or ponding of water at the interface.

The test material (tire shreds or gravel) was placed inside the form and sand was simultaneously placed and compacted around the outside of the form in each pit. The tire shreds were placed in lifts, and each lift was compacted by using a hand tamper. There was noticeable compression of the tire shreds upon compaction. The thickness of each lift after compaction was about 8 inches. The gravel in the control trench was also placed in lifts and compacted with the tamper, but there was negligible compression during compaction of the gravel. After the 2-foot thickness of test material was in place, the form was removed from the trench. A 4-inch diameter perforated PVC pipe was then placed on top of the 2-foot thick layer of test material to distribute the effluent. A thin layer of material (tire shreds or gravel) was placed on top of the effluent pipe, and then the test trench was covered with a geotextile filter. Sand was then placed and compacted around the sides of the trench and above the geotextile, to bring the grade up to the surrounding ground surface elevation. Minimal compaction was applied to the sand placed above the 2-foot wide test trench, to preclude potential problems due to compression of the underlying tire shreds once the effluent pipe was in place.

RESULTS

After the installation of the leaching-field trenches, each trench was dosed with of septic tank effluent. The septic tank effluent was piped to a Distribution box (D-box) and the D-box was run on a timed cycle. Approximately every fourteen days, samples of effluent from each of the three leaching-field trenches and a sample of D-box wastewater (which is the influent to the three trenches) were taken and analyzed for the following parameters. All the analyses were conducted according to the procedures detailed in the Standard Methods [9]. For metals determination, each sample was digested with concentrated nitric acid before analysis.

pH PROFILE

It may be noted from *Figure 2* that the pH of the effluent from the leaching-field trenches closely followed the pH of the influent (D-box) for January –

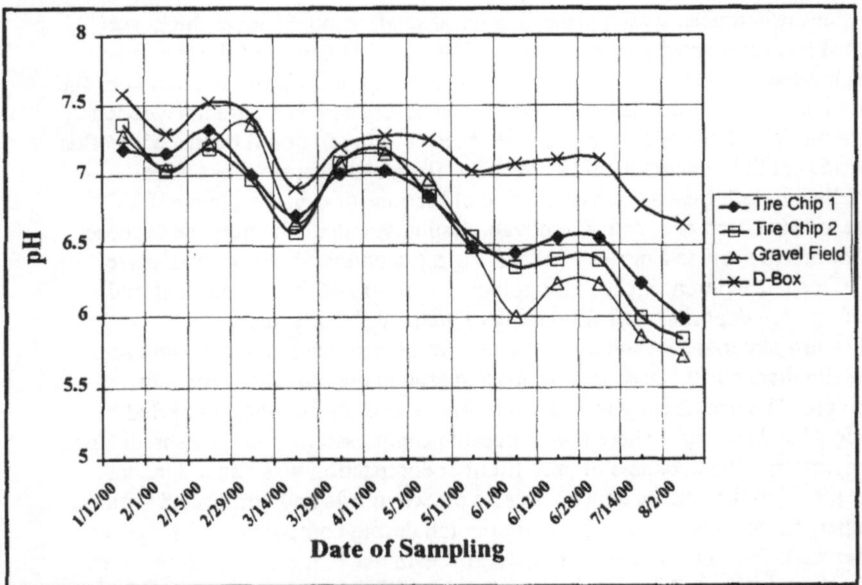

Figure 2: pH Profile of the Effluent from the Three Leaching - Field Trenches and the Distribution Box

April 2000 after which the effluent pH was much lower than the influent value. This is because of a) carbonaceous oxidation which produces CO_2 (g), the dissolution of which generates carbonic acid which is a source of H^+, and b) nitrification which consumes alkalinity and thus results in drop in pH.

TOTAL ALKALINITY PROFILE

Alkalinity is a measure of the solution to resist change in pH upon addition of H^- or OH^- ions. In typical wastewater treatment systems, the carbonate system is mainly responsible for providing alkalinity or buffering capacity. The mathematical definition of Total Alkalinity is:

$$TALK = [HCO_3^-] + 2[CO_3^{2-}] + [OH^-] + [H^+] \quad (1)$$

but the $pK_{a,2}$ of the HCO_3^- - CO_3^{2-} system is 10.33. Therefore, at a pH of near neutral, which is the situation with all the leaching-field trenches, the concentration of carbonate is practically negligible. Moreover, the concentration of H^+ or OH^- ions is in the range of 10^{-6} or less. Therefore, under such conditions, the definition of total alkalinity is simplified to:

$$TALK = HCO_3^- \quad (2)$$

After 4/11/2000, the total alkalinity of the effluent from all the three leaching-field trenches dropped significantly (*Figure 3*) vis-à-vis that of the

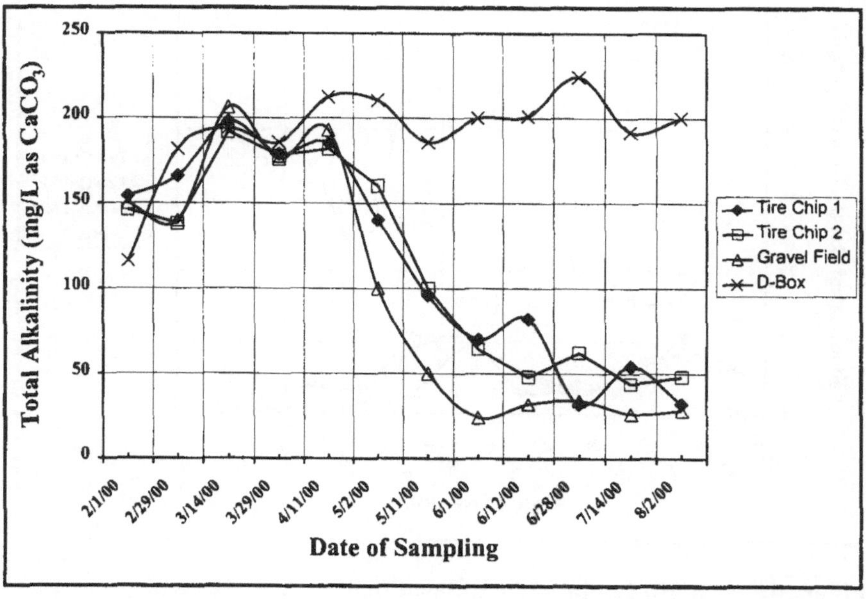

Figure 3: Total Alkalinity Profile of the Effluent from the Three Leaching - Field Trenches and the Distribution Box

D-box influent. The most probable cause of this is microbially-mediated nitrification, as is explained later. This sharp reduction of alkalinity also affects the effluent pH and may exert strong influence on leaching of metals from the surface of the tire shreds.

BOD_5 PROFILE

BOD is a very important wastewater quality parameter. Even though Title 5 Standards do not explicitly specify BOD_5 levels of any leaching-field effluent, the implication is that a well-functioning Title 5 system is capable of reducing the BOD_5 of the domestic wastewater to a level such that it does not adversely impact the quality of the groundwater that it meets. The general consensus is that a leaching-field effluent with a BOD_5 of < 25 mg/L is considered a safe discharge. One of the major goals of this project was to evaluate if tire-shreds, when used instead of gravel in subsurface residential wastewater treatment systems, can provide the same level of BOD_5 reduction in the effluent. From *Figure 4* it is clear that after about five months of operation, leaching-field trenches that use tire shreds provide the same level of BOD_5 reduction as conventional (gravel) systems and that the BOD_5 of the effluent is well below the 25 mg/L recommended limit in both the cases.

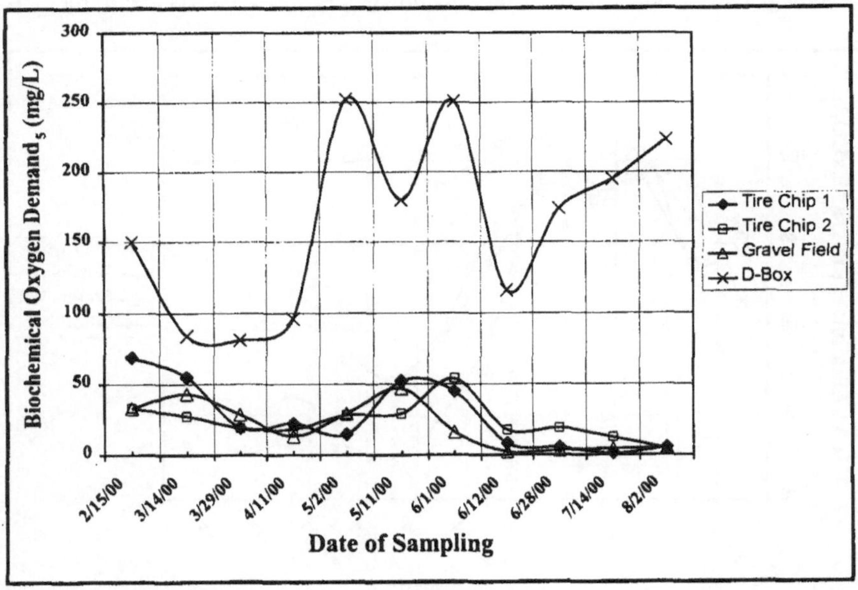

Figure 4: **BOD_5 Profile of the Effluent from the Three Leaching - Field Trenches and the Distribution Box**

FECAL COLIFORM PROFILE

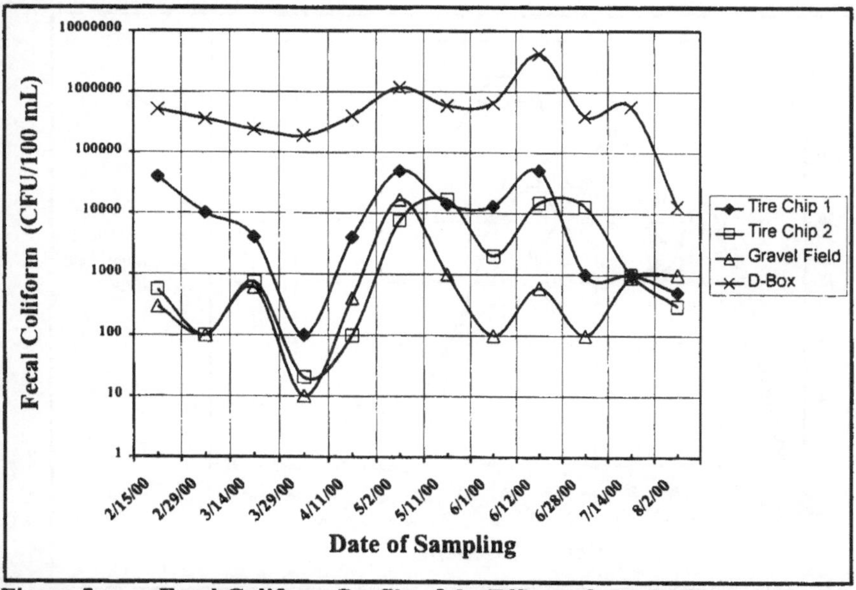

Figure 5: **Fecal Coliform Profile of the Effluent from the Three Leaching - Field Trenches and the Distribution Box**

The coliform group consists of several genera of bacteria belonging to the family Enterobacteriaceae and this group is used as a surrogate for the degree of bacterial contamination of a water (primarily wastewater) source. In this group of coliform bacteria, there is a subgroup of microorganisms that is discharged in the feces of humans and animals and this subgroup is called "Fecal Coliforms (FC)" and they provide a much better measure of wastewater contamination. Conversely, the reduction in the number of Fecal Coliforms (reported as Colony Forming Units/100 mL) is a good measure of wastewater treatment. Since conventional Title 5 systems provide a high degree of FC reduction, our aim was to compare the performance of tire-shred leaching trench with a conventional leaching trench in terms of FC reduction. As is evident from *Figure 5*, initially (for the first five months of operation), the FC count of the effluent from the tire shreds was higher than that of the conventional Title 5 (gravel) trench. However, after about six months of operation, the tire-shred trench was producing an effluent similar to that from the gravel trench in terms of FC count. Thus, it may be concluded that the tire-shred trenches can provide the same degree of wastewater treatment as that of conventional Title 5 systems after about six months of dosing/operation.

AMMONIUM NITROGEN PROFILE

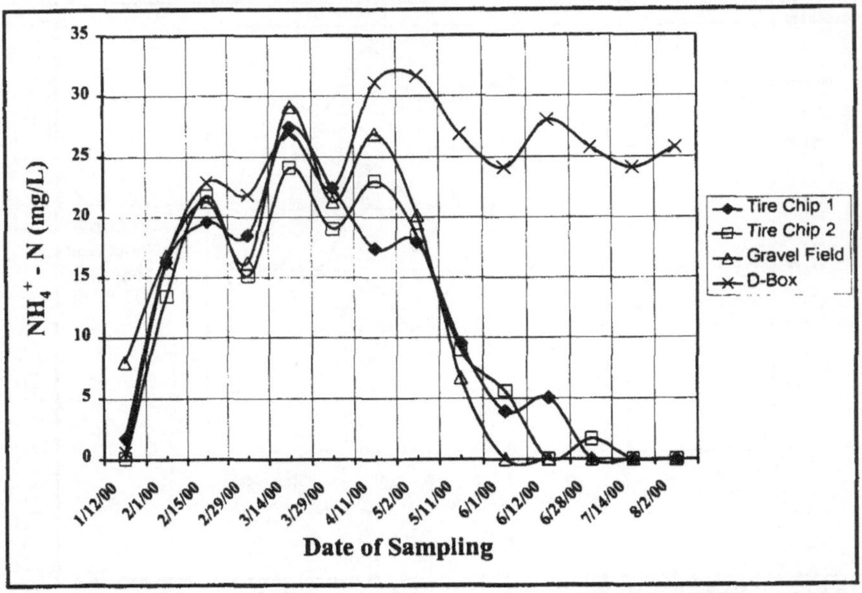

Figure 6: Ammonium Nitrogen (NH_4^+ - N) Profile of the Effluent from the Three Leaching - Field Trenches and the Distribution Box

In fresh wastewater, nitrogen is typically present as organic nitrogen, predominantly in proteins. As the wastewater travels through the septic tank, it is almost completely hydrolyzed to ammonium-nitrogen. Because the septic tank is anaerobic, the reducing environment keeps the nitrogen in the trinegative state. Ammonium ion is a weak acid with an acid dissociation constant, $p_{K,a}$, of 9.4. Thus, at the pH that is typical of septic tank effluent (slightly alkaline), the nitrogen is in the form of NH_4^+. From *Figure 6* it may be observed that the effluent from all the three trenches had an ammonium-nitrogen concentration very close to that of the D-box sample until early May 2000. After this phase, we clearly observe that the D-box continued to have ammonium-nitrogen in the range of 25-30 mg/L as N but the effluent from each of the three leaching fields was almost zero in ammonium-nitrogen concentration. This obviously implies that ammonium-nitrogen is oxidized to nitrate-nitrogen in the leaching field trenches starting early May, 2000. Space limitations do not allow us to validate this hypothesis. We will present evidence of nitrification and phosphorus removal and also show profiles of various metals and common anions during our oral presentation.

CONCLUSIONS

- Tire shreds exhibit high degree of compaction. Therefore, if a tire-shred leaching field is constructed, it should be compacted thoroughly.

- Leaching-field trenches which use tire shreds instead of gravel perform at the same level as Title 5 (gravel) trenches after a stable microbial population is formed but it takes a little longer (almost six months) for the stable biomat to be formed.
- Tire-shred trenches do not leach any toxic metal or inorganic anion or metal that has a secondary drinking water standard MCL at a concentration greater than the MCL, except for Manganese. We found Mn concentration in the effluent from all the trenches (tire-shreds and gravel) at levels higher than the influent and higher than secondary drinking water standard MCL (0.05 mg/L).

REFERENCES

1. Liu, H. S., Mead, J. L., and Stacer, R. G. 1998. *Environmental Impacts of Recycled Rubber in Light Fill Applications: Summary and Evaluation of Existing Literature*, Technical Report #2, Chelsea Center for Recycling and Economic Development, Chelsea, MA, 1998.

2. Humphrey, D. N. 1998. *Highway Applications of Tire Shreds*, Presentation to the New England Transportation Consortium, December 1,1998.

3. Ealding, W. 1992. *Final Report on Leachable Metals in Scrap Tires*, Virginia Department of Transportation.

4. Edil, T. B., and Bosscher, P. J.1992. University of Wisconsin at Madison, *Development of Engineering Criteria for Shredded Waste Tires in Highway Applications*, Wisconsin Department of Transportation.

5. Humphrey, D. N., Katz, L. E., and Blumenthal, M. 1997. *Water Quality Effects of Tire Chip Fills Placed Above the Groundwater Table*, ASTM STP 1275, American Society for Testing and Materials.

6. Minnesota Pollution Control Agency. 1990. *A Report on the Environmental Study of the Use of Shredded Waste Tires for Roadway Subgrade Support*, by Twin City Corp., St. Paul, MN.

7. Zelibor, J. L. 1991. *The RMA TCLP Assessment Project: Leachate from Tire Samples*, Scrap Tire Management Council.

8. 310 CMR 15:00 – The State Environmental Code Title V: *Minimum Requirements for the Subsurface Disposal of Sanitary Sewage*, published by the Commonwealth of Massachusetts, DEP, 1995.

9. *Standard Methods for the Examination of Water and Wastewater*, Published Jointly by AWWA, APHA, and WEF. 2o[th] Edition, 1998.

CONNECTING LARGE SCALE GENERATORS AND RECYCLING MARKETS FOR FOOD RESIDUALS: A HISTORY AND REGULATORY/ECONOMIC/SOCIAL-POLITICAL CASE STUDY FOR NEW JERSEY

PRISCILLA E. HAYES, ESQ., COORDINATOR
New Jersey Solid Waste Policy Group
Department Of Agricultural, Food And Resource Economics
Rutgers University--Cook College
55 Dudley Road
New Brunswick, NJ 08901-8520

DR. DONN A. DERR, CHAIR
Department Of Agricultural, Food And Resource Economics
Rutgers University--Cook College
55 Dudley Road
New Brunswick, NJ 08901-8520

DR. PAULA A. Y. MAAS
Department Of Agricultural, Food And Resource Economics
Rutgers University--Cook College
55 Dudley Road
New Brunswick, NJ 08901-8520

BACKGROUND

Estimates of total food residuals which enter the solid waste stream vary greatly, although residential and non-residential portions of food residuals are roughly equal. Cynthia Greene of the United States Environmental Protection Agency (USEPA) reports that food residuals currently constitute 21.9 million of the roughly 217 million tons of solid waste nationwide, or approximately 10.1%. Of the food residuals generated in America, reportedly only 569 thousand tons of food residuals--2.6% of the total--was diverted to recycling options such as composting or animal feeding.

Using 1999 figures, the New Jersey Department of Environmental Protection (NJDEP) estimates food residuals to be only 4.87% of the total solid waste generated, and on the basis of that percentage and reporting by towns of food residuals recycling undertaken within their boundaries by businesses and residents, concludes that 36.9% were recycled.[1] Of the 304,679.74 tons of food residuals which New Jersey municipalities reported as being recycled within their boundaries, 31,131.07 were sent to swine food residuals feeding operations. Other commercial food residuals recycling was reported as 267,253.21 tons, and residential as 6,295.46 tons.

A third estimate, albeit local, was done in 1992 in Cape May County. The four-season waste composition study, apparently based on actual sorting of garbage, indicated that food residuals comprised 15.8% of the overall waste stream—indeed the author suggests that the percentage may be even higher because of two factors: 1) food residues left in pizza boxes and other containers, and 2) the tendency of those actually sorting solid waste to shy away from food residuals due to its unpleasantness, and thus to underreport it.[2]

A 1996 estimate by Dr. Melvin Finstein, Department of Environmental Sciences, Cook College, Rutgers the State University of New Jersey, indicates that food residuals generated in Mercer County comprise roughly 10% of the overall waste stream—and 19% of the waste not being recycled—but Dr. Finstein cites the Rugg article for support that food residuals comprise a potentially much higher percentage of the waste stream.

This paper summarizes some past studies of the potential for growth in the New Jersey food residuals recycling industry, and briefly describes the history of two forms of recycling within the state, animal feeding and composting. Case studies are presented for these two forms of recycling. The paper's conclusion suggests some direction for action in the state, in order to further encourage this recycling.

STUDIES OF POTENTIAL IN THE NEW JERSEY FOOD RESIDUALS INDUSTRY

In December 1984, a comprehensive study of the New Jersey food residuals feeding industry entitled *Feasibility of Food Waste Recycling in New Jersey: Final Report to the Office of Recycling,*[3] was presented to the NJDEP by researchers at Cook College, Rutgers, the State University of New Jersey. It contained a series of recommendations to revitalize the industry: (1) local regulatory relief; (2) tax relief on farm buildings; (3) listing of food residuals generators and food residuals feeders; (4) hiring of outreach personnel with a background in swine production to assist with animal nutrition, housing and health management; (5) increased food residuals supplies through mandatory recycling; (6) institutional support at the municipal/county/state level; (7) creation of a food residuals recycling coordinating council; (8) financial assistance in the form of grants and low interest loans to upgrade facilities; and (9) research and development to expand the food residuals market including drying and pelletization.

These recommendations were largely ignored by the state government, ironically at a time when public policy was encouraging waste recycling including food waste. Licensed feeders indicated that they could use an additional 600 tons per week over and above the 1,500 tons they were currently feeding in 1984. This would translate into significant waste that would not go to the landfill. Not only would this have served as a savings to the taxpayer in waste management costs, but as a benefit for the rural community in the form of employment opportunities.

In 1996, Dr. Melvin Finstein responded to a request by anti-incinerator activists in the Borough of Princeton for a presentation to the Borough Council on recycling of food residuals. Dr. Finstein's presentation, entitled "Feeding Mercer County's Food residuals To Animals: An Approach That Is In Place, Lowest In Cost, And Best For The Environment," noted that of the 32,000 tons of food residuals generated annually in Mercer County, approximately 1000 tons were being picked up and fed by pig farmers in Gloucester and Burlington counties. Dr. Finstein noted that the cost of this service was half that of traditional waste pick up for disposal in a landfill or incinerator—and opined that much more of the commercial food residuals could be accommodated through animal feeding markets.

THE ANIMAL FEEDING NICHE

The utilization of swine for food residuals recycling has been a time honored practice used for thousands of years. This occurs in many forms—from using a swine herd to utilize incidental food residuals from cooking and canning and freezing, to maintaining a few animals for eventual home use, to food-residuals-feeding farms comprised of hundreds of pigs as a commercial operation.

The first pig farming in the area was in New York City, where pigs roamed the streets eating whatever food residuals had been dumped there, and generating complaints as early as 1720. When the Holland Tunnel opened in 1927, pig farmers were profitably able to locate in the Meadowlands/Secaucus area, because they could go to Chinatown and Restaurant Row in New York City each day and pick up food residuals to feed to their herd, obviating feed costs. At its height the pig industry in this area may have numbered 40,000 animals. However, the pig farmers began to be displaced in 1959 with the construction of the New Jersey Turnpike, through 1976, when Giants Stadium opened and the odor of the remaining pig farms made attendance at Giants games unpleasant. Many relocated to southern New Jersey, particularly to Gloucester County. This location, in turn, has been supported by food residuals from Philadelphia. Unfortunately these residuals were sent with a high percentage of inedible material, thus creating a solid waste management issue on several farms during the 1990s, due to the inedible materials ending up in the manures and littering the farms and surrounding areas.

Due to a number of contributing factors, however, the food residuals feeding industry has been on a steady and sharp decline in the number of operations and the number of swine. The number of pig farms licensed as "garbage feeders" has dropped from 250 in 1963 to 24 today (of the roughly 400 pig farms remaining in the state).[4] It is likely that the growing international concern over disease outbreaks in Europe, such as spongiform encephalopathies (e.g. BSE) and Foot-and Mouth Disease (FMD) will be a major contributor to additional decline in the industry.

The quality of food residuals for feeding does vary depending on its nonedible/nonfood content, water content, and the food residuals source. The very best food residuals are those which are blended, cooked, dried, and pelletized, resulting in a high valued product with a long shelf life. Processed bakery waste (pulverized, dried, and wrappers removed) is also a very high quality product normally used bulked (not packaged). Unprocessed bakery waste includes items that do not meet quality control standards (baking quality or freshness or appearance). The wrappers are often not removed from the food residuals, thus generating a nonbiodegradable solid waste problem that becomes mixed with the soil and animal waste. A very broad category of food residuals is processing/distribution sources. This includes residue from food processing plants, grading and packaging facilities, and fruit and vegetables damaged en route to the final destination. Food residuals from restaurants, catering halls, institutions, corporate and school/college cafeterias may be collected as separated and pulped (removal of excess moisture) or separated but not pulped. Food residuals collected from such sources may be a mixture of prep waste and plate waste, or simply one or the other. The lowest quality food residuals include unseparated and municipal waste with a high percentage of nonfood items that accumulate at the food residuals facility and ultimately must be collected and transported to the municipal solid waste facility.

WILENTA CASE STUDY

Wilenta was one of the pig farms which was displaced from the Meadowlands area where it had been hauling waste to its pigs from such New York City restaurants as the Waldorf-Astoria and Mama Leone's. Although it maintains its offices and facilities for consolidating food residuals in Secaucus, the actual pig farm is now located in Lakewood, New Jersey. Wilenta is run by two Wilenta brothers, one of whom owns the pig farm and the other the trucking company. Many of their current customers are processing firms; bakery waste provides especially good feed for the pigs.

Up until two years ago, Wilenta was not actually charging many of its customers to haul waste, which caused the business to face severe financial strain. While Wilenta maintains rates below those of the prevailing tipping fee in any given area, their profit is made on the hauling end, since the pork market has not been particularly vital in recent years.

Wilenta will only service generators who are willing to let Wilenta do the hauling themselves, since otherwise their chief source of revenue disappears. They work as a niche market, in part, because the processing required for feeding to the pigs is relatively minimal or not required at all (produce, bakery and sweet waste), and thus, their costs are relatively low, offset somewhat by the sale of animals. They fear that any change in processing requirements will increase costs, and reduce their relatively slim profit margin.

ENVIROFEED CASE STUDY

In the mid 1990s, Enviro-Pak was formed with the purpose to "recycle heavy food residuals from restaurants."[5] In 1995, a swine feeding program was established; from 1996-1998 a collection system was developed and expanded, with food still going to a swine farm, and with such customers as Princeton University signing on. At the same time, planning began for a feed plant. By 1998 a Research Development and Demonstration permit had been obtained from NJDEP for a plant to be called EnviroFeed, and work was well along on siting and outfitting the plant. Full production was to begin in 1999.

The collection system created, utilizing MacNeil Trucking, involves sorting of food residuals into 32-gallon toters, with or without plastic liners, by trained employees at each generating facility. The small size of the containers and the ability to dump them manually into the back of a standard waste truck allows the MacNeil employees to remove any non-food residuals during this process. The end result is extremely clean and highly desirable food residuals. Collection from each generator occurs either every two or three days. Collected food residuals were tipped out at the plant, and directed onto a conveyer belt, where any remaining non-food contaminants are picked out. Magnets and other equipment were developed to remove further contamination such as silverware. The cleaned materials were mixed and pulped, then wheat middlings were added, to reduce overall moisture levels, which could be as high as 76%. From there, the mixture was pressed out into long ropes of feed, and entered a dehydrator. In addition to removing remaining moisture, this served to heat the feed to the high temperature meeting the requirements of the United States Department of Agriculture.

One of EnviroFeed's most satisfied customers was the Newark Airport Hilton. Marco Foelske, Director of Food and Beverage there, had been looking for a way to avoid having the smelly and leaking dumpsters which were created when food residuals was mixed with the hotel's other trash. When approached by John Metzger, salesman for EnviroFeed, he was immediately attracted to the EnviroFeed collection system, which helped him maintain the clean environment which he felt was so necessary to the hotel's image and business. He diverted not only food from the restaurant areas, but from room service tray returns. The stewarding department was in charge of checking employee performance on a daily basis. In Newark, Mr. Foelske was able to implement the system without extra labor costs; at the Hilton in Short Hills, an extra person—a recycling manager—was hired. Mr. Foelske estimated his cost savings as $250 per month, since EnviroFeed charged less than he had been paying to dispose of the food residuals along with his other waste.

While EnviroFeed was careful to set collection and hauling rates which would make disposal of food residuals with them cheaper than with a conventional hauler, there were other advantages besides the economic benefits to generators who were recycling food residuals with EnviroFeed. As with the Hilton in Newark, there were cleanliness advantages, and because food residuals sat for less time, flies did not build up. The wheeled carts did not require employees to lift of heavy bags into a dumpster, meaning a decreased risk of back injuries.

Market and other conditions proved increasingly unfavorable to EnviroFeed. A great deal of trial and error was required to develop the equipment to properly sort and process the food residuals, and the wheat middlings were an expensive additive. Odor complaints from neighbors led to an air permit which required construction of a taller stack. On the feed end, the material, which was expensive to produce, was competing with feed corn, the market price for which was at a thirty year low (and kept getting lower). Tipping fees remained low, affecting the amount that EnviroFeed could charge the generating facilities for hauling their food residuals. EnviroFeed began to divest itself of its supermarket customers, whose purely vegetative food residuals was typically the highest in moisture content and lowest in protein, both problematic to the feed process (the New Jersey Solid Waste Policy Group, (NJSWPG)[6] was able to partner these supermarkets with Woodhue, using the same hauler and collection system). Recently, Envirofeed discontinued service to its restaurant/food service customers. It is hoping to reopen if market conditions improve.

THE COMPOSTING NICHE

"Composting is a method of solid waste management whereby the organic component of the solid waste stream is biologically decomposed under controlled conditions to a state in which it can be handled, stored, and/or applied to land without adversely affecting the environment."[7] The finished product, compost, is generally used as a soil amendment; normally compost does not contain levels of nitrogen, phosphorus and potassium to qualify as a fertilizer. Unlike most chemical fertilizers, however, compost contains at least small quantities of many other beneficial nutrients, as well as significant amounts of organic material, both these characteristics make it very valuable to soil and plant health. Composting is normally done outdoors, in piles or windrows which are turned regularly to promote aeration and reduce built-up heat. Anything which was alive or which is a waste from a living organism can be composted; normally outdoor composting is limited to vegetative wastes and manures from plant eating animals, since animal products, including dairy products, tend to attract rodents and flies. These materials can be composted in vessel.

Although backyard composting, including both vegetative food residuals and yard waste, has of course been practiced for many years across the state, large scale composting by municipalities or commercial operations has been limited until recently to yard waste—i.e. grass clippings, leaves, yard trimmings, etc. The first large scale (non-residential) food residuals composting in the state was performed at American Soil Inc. (ASI) in the early 1990s.

ASI, founded in 1986, had been permitted since 1987 to accept leaves and grass. Initially they applied for a Research Development and Demonstration permit from DEP to begin composting food residuals. They spent at least 5 months doing advance education on food residuals composting, and the pilot project lasted six months. Since the program had been a success, ASI applied for a permanent permit, which DEP initially refused. A full food residuals permit was issued in 1993; ASI used it to compost vegetative waste and cardboard from supermarkets, several vendors from the Paterson Farmers' Market and occasional vegetative food residuals from the United States Custom House.

WOODHUE CASE STUDY

Woodhue Farms consists of two adjoining composting sites, the Woodhue Ltd. site and the Bryony Limited Corp. site, in Wrightstown, New Jersey. Since 1980, they have been permitted to compost food processing waste. By 1986, they were processing all Ocean Spray processing residues amounting to about 20,000 tons per year. Later Woodhue was permitted to accept vegetative waste and crop residues, biosolids, and potable water residuals. In 1996, three years after ASI had been granted the first permit to accept and compost supermarket waste, Woodhue asked to be permitted, and received a permit for the same material. This was just before the solid waste regulations changed and required a new permit, i.e. a Class C permit. Woodhue's application for this permit was administratively complete in January 1998. As of this date, the permit has still not been issued.

Woodhue's grandfathered permit allows them to compost only vegetative and bread-based food residuals, not any meat or dairy residuals. Woodhue is also permitted to compost cardboard along with the food residuals, including waxed cardboard. Composting takes place on the ground in windrows turned periodically by a Wildcat compost turner.

The Woodhue-Bryony sites have only one neighbor, a woman who raises sheep, on the northwestern corner of the site. This neighbor has made repeated complaints about the site, and as a result the NJDEP has classified her property as a sensitive land use, requiring a buffer of 1000 feet for food residuals and grass composting.

In 2000, American Soil ceased taking supermarket waste; about the same time EnviroFeed began to phase out its supermarket waste collection. The NJSWPG was instrumental in arranging partnerships which brought much of this waste to Woodhue, and brought Woodhue the collection services of MacNeil Trucking. The collection system was the one described above with 32 gallon toters, created by EnviroFeed and MacNeil Trucking. As noted, the end result is extremely clean and highly desirable food residuals, which Woodhue contrasts to other residuals collected in a compactor, which contain much more contamination.

Up until June 2000, Woodhue had established collection from only about 30 to 35 supermarkets, but the addition of the supermarkets previously serviced by EnviroFeed and American Soil brought the total to about 125. Woodhue is currently phasing in supermarkets from several new chains, as fast as personnel at each supermarket can be trained to properly sort the appropriate food residuals into collection containers.

In November 1999, in anticipation of current DEP regulations sunsetting in 2001, members of the NJSWPG met with Tim Bartle and Kevin Sullivan of DEP to discuss and suggest possible changes to the regulations as they affect food residuals diversion. At that time, a variety of enforcement issues were raised, chiefly involving composting facilities. Accordingly, Tim Bartle suggested that the NJSWPG and food residuals recyclers meet with a new assistant commissioner in charge of these enforcement issues, Cathy Tormey.

At the February 10, 2000 meeting with members of Ms. Tormey's staff, DEP enforcement personnel confirmed that DEP had placed composting facilities under increased scrutiny, starting in 1999. This was due to a documented rise in the level of violations for municipal composting facilities for the previous year, which DEP felt pointed to a need to do monthly inspections. Monthly inspections and increased scrutiny have also lead to an increased issuance of Notices of Violation (NOVs). In the past, inspectors might simply note a problem/potential violation on an inspection report, rather than issuing a NOV. This had led to problems with follow up, so the practice of simply making notations has been discontinued, and NOVs are being issued.

Woodhue began to receive this increased level of scrutiny, along with other composters, and had a series of NOVs involving slight windrow width differences and ponded water immediately after a rainstorm. Woodhue challenged each NOV, but was concerned that regardless of its challenge, and any changes made as a result thereof, that an enforcement file was building up. Enforcement personnel suggested that such NOVs might point to a need to change permit language; however the Class C application was still pending, and there seemed no immediate opportunity to do so.

In meetings held between DEP personnel, Woodhue representatives and a representative of the New Jersey Office of Sustainable Business in early fall 2000, Woodhue described the pattern of small NOVs and was urged to apply for the Silver Track Program, a program which is meant to signal a departure from the traditional regulatory approach and offer participants additional flexibility in managing their operations and an opportunity to implement eco-efficient strategies and improve relations with communities and with regulators. The review of permit applications and renewals of Silver Track participants will be coordinated by a single point of contact within the Department, and permit processing was to be expedited. Woodhue entered the application process for the Silver Track in an effort to avoid future build up of an enforcement record similar to the past series of NOVs it had objected to. An enforcement representative blocked the application by citing the very NOVs which had been the reason for DEP representatives suggesting application in the first place.

Currently, Woodhue still does not have a Class C permit, and has been advised that it must resolve county plan issues before final action can be taken. In the meantime, Woodhue continues to expand its customer base under its grandfathered approval, and is pursuing innovative new services and recycling opportunities which can be offered at its site.

CONCLUSIONS

Recycling food residuals by direct feeding to animals, particularly pigs, continues to be a niche which works economically because it provides the farmer with free feed—indeed over time, the farmers have begun to charge a tipping fee to cover collection and other costs. Moreover, only minimal processing is required, normally cooking to a certain temperature for a specific period of time to meet the requirements of the Swine Health Protection Act,[8] if the residuals include meat or other processed foods; no processing at all, if the residuals include only vegetative and bread or candy waste. Recently, the practice of feeding food residuals has been called into question by the potential for transmission of spongiform encephalopathies and FMD which affect animals through feeding of certain animal parts or feeding residuals containing meat without proper processing. There is a nationwide effort underway to determine whether any changes should be made to the federal regulations which govern the conditions for feeding food residuals to animals.

Pelletizing, which could represent the future of animal feeding, because of its creation of a product which is readily transportable and has a long shelf life, has not been feasible due to market conditions and lack of legal infrastructure or incentives to support it. Composting exists in New Jersey largely through the tenacity of one private firm, and through its association with the NJSWPG. Like direct feeding of food residuals to animals, this requires relatively minimal processing. The food residuals feeding industry, in particular, is at the mercry of the marketplace for grains and meat prices, as well as government regulations. In order for the food residuals recycling industry to remain economically viable, there must be a well defined public policy, including mandatory recycling for the larger food residuals generators, local/county/state institutional support, and financial assistance in the form of grants and low interest loans to both generators and recyclers. The case study accounts here reflect the most innovative entrepreneurs in the state, and if they do not prevail economically, then attrition will continue, until there is no longer any food residuals recycling in New Jersey.

---

[1] 1998 New Jersey Material Specific Recycling Rates, with Note: "Waste stream composition analysis performed in 1990. Subsequent changes to the waste stream (primarily as a result of increases in reported recycling tonnages) have not been taken into account, which have affected the 'Total Tons of Solid Waste Generated' and '% of Solid Waste Material Recycled' numbers for specific materials listed, though not for the 'Totals' at the bottom of the table. The waste composition study will be revised for future years to account for this anomaly."

[2] Rugg, Mack, and Hanna, Nabil K. "Metals Concentrations in Compostables and Noncompostables of Municipal Solid Waste in Cape May County, New Jersey." in *Composting Frontiers, Vol. 1, No. 1.*, Susan Mazzocchi, Editor. Pp. 4-10.

[3] Suhr, Jeffrey L., Higgins, Andrew J., Price, A. Turner, and Derr, Donn A. December 1984. "Feasibility of Food Waste Recycling in New Jersey: Final Report to the Office of Recycling," Cook College, Rutgers, the State University of New Jersey, New Brunswick.

[4] There are also several farms in the state which feed food residuals to other animals, notably cattle and goats.

[5] From EnviroFeed materials in preparation for SWAC presentation.

[6]The NJSWPG is an initiative designed to coordinate resources at Rutgers University and across the State to address real life solid waste issues. The NJSWPG was created to identify actions to help prevent future waste problems that may negatively impact on the quality of life, and to inform New Jersey's solid waste decision-makers about policy options for enhanced solid waste management, in both environmental and economic senses. Using Rutgers as a base provides a neutral forum for policy visioning and for coordinated research by technical and policy experts, economists, social scientists and stakeholders.

[7] Golueke, C.G. (1977). *Biological Reclamation of Solid Waste.* Emmaus, PA, Rodale Press.

[8] Swine Health Protection Act, 7 U.S.C. § 69 (1998).

A RE-USABLE LANDFILL : THE AEROBIC BIOCONVERSION CELL

Nickolas J. Themelis[1] and Young Hwan Kim[1,2]
[1]Department of Earth and Environmental Engineering, Columbia University;
[2]Department of Materials Science, Hong-Ik University, Seoul

ABSTRACT

Land that is used for disposing biodegradable materials such as food, plant and paper wastes is effectively put out of use forever. A city of 150,000 effectively wastes about five acres per year to dispose its wastes in what is called, euphemistically, a "landfill". In addition, such a landfill remains for centuries after closure a source of potent greenhouse gases (methane, etc.) and leachate solutions. This paper will discuss the design of a re-usable landfill: The Aerobic Bioconversion Cell. The preferred values of operating parameters and of the biochemical rate constants of oxidation to CO_2 and H_2O were identified. Energy and material balances were then constructed for a large, 3-m deep aerobic cell holding 1,440 tons of the 'wet' component of organic wastes (major organic constituent: $[C_6H_{10}O_4]_n$). It was found that conduction/convection and radiation losses to the surroundings amount to a relatively small fraction of the chemical heat released by oxidation. The excess heat must be removed by means of an upward water-saturated air flow that is several-fold the stoichiometric requirement for oxidation. From a process engineering viewpoint, the principal difference between anaerobic and aerobic biodegradation is that in the former much of the chemical heat contained in the wastes is transformed to methane and, if collected, can be used as a fuel elsewhere. In the case of aerobic oxidation in a packed bed reactor, this heat is released by oxidation and need to be removed by a very large flow of air through the bed.

Separation of Municipal Solid Wastes to "wet" and "dry" streams

The municipal solid wastes (MSW) collected in the U.S. amount to 190 million tons (209 million short tons) per year. Excluding high-density materials like metals and glass, MSW consist of food and plant wastes, paper, plastics, wood, leather and various fabrics. The unpleasant odors and liquids associated with "garbage" are due to the decomposition, in the absence of an adequate supply of oxygen, of the putrescible organic components of food and plant wastes. These organic materials contain 50-70% water, an essential element for bacterial action. When they are stored for days in containers, transfer stations and transport trucks they decompose emitting unpleasant gaseous and liquid emissions. In warm weather, the rates of bioconversion increase.

About 60% of the U.S. MSW, i.e. an estimated 127 million tons (140 million short tons) per year (Franklin Associates, 1995), is disposed in landfills that are effectively sealed so that bioconversion reactions catalyzed by bacterial organisms are carried out in the absence of oxygen and produce principally methane (natural gas) and carbon dioxide. In contrast, most of the bioconversion organic matter in nature, e.g. in fields and forests, occurs in the presence of ·ygen, i.e. aerobically, and the products are water and carbon dioxide.

As available sites for conventional landfilling become scarce, as is already e case for New York City, it makes sense to separate the "wet" fraction (plant ıd animal organic matter) from the "dry" fraction of the MSW (paper, plastics, etal, glass, etc.) at the source, i.e. at the household, and treat each fraction ›parately. This is already done in at least two Canadian communities, Halifax, ova Scotia, and Guelph, Ontario www.columbia.educu/earth). The Earth ngineering Center of Columbia University is collaborating with Rutgers University examining the possibility of New York citizens separating at the source the SW into a "wet" stream that can be composted. The "wet" fraction, that in NYC ›nstitutes less than 30% of the total MSW (i.e. about one million tons per year), ecomposes quickly and if it is commingled with the rest of the MSW complicates ansport and processing. This presentation will examine the material and thermal nergy balances for a large size cell where food and plant wastes are composted erobically by means of forced air injection.

omposition of MSW

Tchobanoglous et al (1993) estimated that the U.S. MSW contains from 6-8% food wastes and 5-20% yard wastes. The "typical U.S." MSW was estimated › contain 9% of food wastes and 18.5% yard wastes (Table 1); these numbers re not much different from a New York City study (SCS Engineers, 1992) that stimated that the food wastes represented 12 % of the city's MSW and the total rganic wastes less than 30%.

Table 1. Composition of various components of the MSW
(Tchobanoglous, Theisen, and Vigil, 1993

	Moisture	Volatile matter	Fixed carbon	Non-combustible	kJ/kg* as collected
lixed food wastes	70	21.4	3.6	5	4180
'ruit wastes	78.7	16.6	4	0.7	3970
leat wastes	38.8	56.4	1.8	3.1	17730
lixed paper	10.2	75.9	8.4	5.4	15600
'ard wastes	60	30	9.5	0.5	6050
;ommingled Residential MSW	21	52	7	20	11630

The ultimate (atomic) analysis of the various types of wastes is shown in 'able 2 (Tchobanoglous et al, 1993).

Table 2. Ultimate analysis of various components of the MSW
(Tchobanoglous, Theisen, and Vigil, 1993)

	Carbon	Hydrogen	Oxygen	Nitrogen	Sulfur
Mixed food wastes	48	6.4	37.6	2.6	0.4
Fruit wastes	48.5	6.2	39.5	1.4	0.2
Meat wastes	59.6	9.4	24.7	1.2	0.2
Mixed paper	43.4	5.8	44.3	0.3	0.2
Yard wastes	50.1	6.4	42.3	0.1	0.1

On the basis of the composition data of Table 2 and the atomic weights of the respective elements, the composite molecular formula corresponding to each material component of MSW is calculated to be as follows:

Mixed food wastes: $C_6H_{9.6}O_{3.5}N_{0.28}\,S_{0.2}$
Mixed paper: $C_6H_{9.6}O_{4.6}N_{0.036}\,S_{0.01}$

The above formulae are useful in determining the principal reactions that occur during bioconversion, the reaction products, and the expected heats of reaction. It can be seen that sulfur and nitrogen are relatively minor components and occur principally in mixed food wastes. The structure of mixed food and plant wastes can be approximated by the molecular composition $(C_6H_{10}O_4)_x$, which, for x=1 corresponds to that of at least ten organic compounds (Table 3, after Roinen, 1999). The most common value for the heat of formation of these compounds is near -230 kcal/mol. All thermodynamic calculations in this study utilized the commercially available chemical thermodynamics program HSC Chemistry 4.0 (Roinen, 1999).

Anaerobic bioconversion of MSW

The principal products of anaerobic decomposition are methane and carbon dioxide; ammonia, hydrogen sulfide and mercaptans (sulfonated hydrocarbons) are also generated. In the absence of adequate moisture content throughout the reactive landfill, the rate of bioconversion of wastes is very slow or can cease altogether. Old landfills have been opened after several decades and little decomposition of the entombed material has been observed at locations where there was no ingress of moisture. On the other hand, if a large amount of precipitation water is allowed to permeate through the landfill, the leachate solutions formed within it must be collected and treated continuously so they do not contaminate the groundwater and adjacent bodies of water.

Table 3. Identified organic compounds with formula $C_6H_{10}O_4$ (mol. weight: 146.143 g/mol; after Roinen, 1999)

Chemical formula (Structural formula)	Chemical name (common name)	Heat of formation cal/(mol*K)
$C_6H_{10}O_4$ ($HOOCC(CH_3)_2CH_2COOH$)	2,2-Dimethylbutanedioic acid (2,2-Dimethylsuccinic acid)	-232.316
$C_6H_{10}O_4$ ($(COOCH_2CH_3)_2$)	Diethyl oxalate (Ethanedioic acid diethyl ester)	-176.774
$C_6H_{10}O_4$ $HOOCCH_2CH(CH_2CH_3)COOH$	Ethylbutanedioic acid (Ethylsuccinic acid)	-232.168
$C_6H_{10}O_4$ ($CH_3COOCH_2CH_2OOCCH_3$)	1,2-Ethanediol diacetate (Ethylene glycol diacetate)	-192.832
$C_6H_{10}O_4$ ($(CH_3COO)_2CHCH_3$)	1,1-Ethanediol diacetate (Ethylidene diacetate)	-194.027
$C_6H_{10}O_4$ ($HOOC(CH_2)_4COOH$)	Hexanedioic acid (Adipic acid)	-231.778
$C_6H_{10}O_4$ ($HOOCCH(CH_3)CH(CH_3)COOH$)	meso-2,3-Dimethylbutanedioic acid (meso-2,3-Dimethylsuccinic acid)	-233.599
$C_6H_{10}O_4$ ($HOOCCH(CH_3)CH(CH_3)COOH$)	Racemic-2.3-Dimethylbutanoic acid (Rasemic-2,3-Dimethylsuccinic a.)	-233.599

The quantity of natural gas generated during anaerobic decomposition can be determined by considering the simplified molecular formula of organic waste derived earlier and assuming that the decomposition reaction proceeds as follows:

$$(C_6H_{10}O_4)_x + 1.5H_2O = (C_6H_{10}O_4)_{x-1} + 3.25CH_4 + 2.75CO_2 \qquad (1)$$

Depending on the assumed heat of formation of the complex organic compound (Table 3), this reaction ranges from slightly endothermic to slightly exothermic; generally, in landfills there is some ingress of oxygen that generates heat by means of partial aerobic reaction. According to Equation (1), the product gas contains about 54% methane and 46% carbon dioxide. It also contains water vapor at the saturation pressure corresponding to the landfill temperature, plus small amounts of ammonia, hydrogen sulfide and other minor constituents. If it is assumed that the $(C_6H_{10}O_4)_x$ component is 25% by weight of the MSW stream, the maximum amount of natural gas that can be produced from MSW is calculated from Equation (1) to be 134 standard m^3/metric ton (4,300 standard cubic feet per short ton). However, as the hydrocarbon chains become shorter, the organic molecules become more stable and the rate of bioconversion decreases substantially. The reported (DOE/SRI study: see Franklin, 1995) maximum capacity of MSW in landfills to produce methane is 62 standard m^3/t (2,200 scf/short ton) of CH_4, i.e. 50% of the total amount projected by Equation 1.

The overall capture of methane gas produced in U.S. landfills has been estimated at about 66% (Dennison, 1996). On the basis that the total U.S. MSW is 190 million tons (209 million short tons), 61% of it is landfilled, and 66% of the generated methane is recovered, the carbon loss to the atmosphere in the form of CH_4 is calculated to be 1.45 million tons C/year. Since methane is 21 times more potent as a greenhouse gas than CO_2, the methane losses from landfills may correspond to 36.4 million tons of carbon, or about 2% of the total U.S. contribution to global warming.

Aerobic bioconversion of organic wastes

In the presence of an adequate supply of oxygen, moisture, and aerobic bacteria, the hydrocarbons contained in organic wastes undergo partial oxidation to smaller molecules. This results in the destruction of odor- forming compounds and the formation of carbon dioxide, water vapor and a compost product that can be used as soil conditioner. In its natural form, aerobic bioconversion of organic matter occurs naturally beneath the surface in fields and forests.

In order to divert organic wastes from landfills, many communities and private companies operate aerobic composting plants. These range from "windrow" composting, to sophisticated "in-vessel" bioreactors, such as the Bedminster (1994) bioreactor, a rotating cylindrical vessel through which the raw organic material flows countercurrently to an air flow. Haug (1993) describes several other types of operating aerated bioreactors. In the U.S., aerobic composting amounts to about 6.3 million short tons/year, i.e. 3.5% of total MSW.

By representing organic wastes by the simplified chemical formula presented earlier, the aerobic bioconversion reaction can be expressed as follows:

$$(C_6H_{10}O_4)_x + 6.5O_2 = (C_6H_{10}O_4)_{x-1} + 6CO_2 + 5H_2O \qquad (2)$$

In contrast to anaerobic bioconversion, this reaction is strongly exothermic; for an assumed heat of formation of the organic matter of 230 kcal/mol, as discussed earlier, Equation 2 generates -616 kcal of heat per mole of organic matters reacted. For comparison, the heat of biodegradation of glucose ($C_6H_{12}O_6$) to carbon dioxide and water is –673 kcal/mol.

A large scale aerobic bioconversion cell

One of the objectives of this study was to compare the thermodynamic heat generated by aerobic bioconversion in a cell with its capacity to dissipate this at. Heat is dissipated in three ways: as latent heat of evaporation of water, as nsible heat in the air flow through the cell, and as heat loss by convection and diation from the surface of the cell exposed to the atmosphere.

The cell in this study is assumed to be 6-meter (20-ft) wide and 3-meter)-ft) deep. Its length would depend on the tonnage of material to be processed, d for the purposes of this study was assumed to be 100 meters (330-ft). Cells of nilar geometry are presently used for aerobic composting (e.g., the Paygro or e Longwood cells) but, in contrast to the static cell examined in this study, the aste material is turned over at frequent intervals by mechanical means, in order expose the wastes to air and water and dissipate the heat generated by the oconversion reaction; the composting process usually lasts three weeks and is llowed by a "curing" period of three to four weeks.

In the hypothetical cell of this study, it is assumed that the packed bed of astes is not disturbed mechanically during the composting period. Air is injected ough a grid floor by means of a ventilation plenum that forces air through the d of wastes; such a device is presently used in over 24 installations of the static nnel" composting system (Haug, 1993); however, a network of perforated pipes the bottom of the cell (Hudgins and March, 1998) can also be used. During eration, adding water at the top of the bed controls the moisture level. At the d of the bioconversion period, a front loader or other device moves the mposted material to the curing area.

From the analysis of mixed food wastes (Table 2), the C/N ratio is lculated to be 18. Here it is assumed that the organic material feed contains out 10% of paper waste (C/N=143), resulting in a C/N ratio of 30.

Tchobanoglous (1993) reports the typical density of food wastes to be 890 /m^3. By mixing food waste with 10% paper waste, the initial moisture content of lk stream is estimated at 65% and its bulk density at 0.8 (800 kg/m^3). Therefore, e loading capacity of our hypothetical cell (1800m^3) is 1,440 metric tons per mposting period.. On the basis of the molecular composition of food wastes, $_{3}H_{9.6}O_{3.5}N_{0.3}$, and the assumption that they consist mainly of carbohydrates, otein and fat, the dry base weight ratio, mole fraction and density are proximated in Table 4.

able 4. Estimation of composition of mixed food and 10% paper waste (dry basis)

	M.W.	wt%	mol. fraction	specific gravity	Heat of oxidation, kcal/kmol
rbohydrates($C_6H_{10}O_5$)	162.142	75.7	0.68	1.27-1.61	-603.0
tein ($C_2H_5O_2N$)	75.067	11.7	0.25	1.14	-163.8
and oil($C_{16}H_{32}O_2$)	256.428	12.6	0.07	0.85	-2387.0
erage of mixture				1.34	-614.6

It can be seen that the computed heat of oxidation of the mix is very close to that calculated on the basis of Equation 2.

Required water flow into the cell

Once the exothermic biodegradation starts, sufficient water flow must be provided over the surface of the cell so as to saturate the air flow and also maintain a 50% moisture content throughout the bed until the end of composting. On the basis of these approximations, the average Free Air Space (FAS) and the porosity of the packed bed are calculated to be 34% and 76.5%, respectively.

Cell airflow, moisture level, and temperature

Keener (1997, 1998) summarized the operating conditions and experimental results of a large number of pilot aerobic bioconversion studies. The preferred aeration rate was expressed in terms of volumetric airflow rate per unit mass of waste and ranged from 0.35 to 0.97 liters/min/kg. Several authors (Bernreuter and Stessel, 2000) have recommended an aeration rate of 0.5 liters/min/kg of waste. It is interesting to note that this rate is nearly the same as the average flow rate of 2,100 scf/h/dry ton, reported by Haug (1993). Vander Gheynst (1997) reported aeration rates of 0.06-0.94 liters/min/kg, desirable moisture content of 45%, and temperature of 55°C. Temperatures much higher than 60°C are considered to be undesirable because they may kill all but the most thermophilic bacteria; also, in the absence of local moisture, high temperatures may result in spontaneous ignition in the bed; in fact, this occurred at times in the 15-m x 150-m x 7-m deep aerobic landfill in California (Merz and Stone, 1962).

The preferred moisture content in the ABC of this study ranges from 45-60% and the cell temperature is maintained below 60°C.

Overall rate of bioconverting - Specific rate constant

One measure of the rate of bioconversion mentioned in the literature is the Specific Oxygen Uptake Rate (SOUR), also called "respiration activity". For example, aerated composting piles were found (Sesay et al, 1998) to have an initial SOUR value of about 19mg O_2/g material per hour; this value dropped to 1.81 after 51 days of aeration. By considering that it takes about 1.425g of oxygen to decompose 1g of organic matter (Equation 2), the SOUR value of 1.81mg O_2/g/h can be converted readily to a more useful measure of the bioconversion rate: The fraction of initially present organic matter that is decomposed per day. Accordingly, the 1.81 mg O_2/g/h reported by Sesay after 51 days corresponds to a fractional degree of reaction of 0.0305/day. Muller et. al. reported similar values (1995; 0.0205/day after 150 days).

On the other hand, Murphy et al (1995) reported lower bioconversion values. In their tests, the fractional rate of decomposition of a MSW/sewage sludge mixture after ninety days, as measured by analyzing for cellulose content, was only 0.0025/day.

An analysis of numerous pilot studies by Keener et al (1998) considers bioconversion to be a first order reaction. The present study also assumes that under controlled moisture, air and temperature conditions, the composting reaction will proceed as a quasi-first order reaction:

$$dc/dt = -k_r c \qquad (3)$$

The k_r values presented by Keener (1998) range from 0.024 to 0.083/day. A study by Hamoda et al (1998) reports higher values ranging from 0.053 to 0.180/day.

After examination of the Schulze and Wiley's experimental results, Haug and Ellsworth (1993) suggested the following equation for the biodegradation of ground MSW mixed with de-watered, digested sludge:

$$k_r = 0.00632(1.066)^{T-20} \quad \text{g BOM oxidized/g OM/day} \qquad (4)$$

where BOM: biodegradable organic matter, and OM: organic matter

Because thermal inactivation occurs above 65°C, the above equation is only valid below 65°C. Moreover, moisture content and available free airspace and oxygen contents will affect the reaction rate also. By counting all these factors, the following expression to describe effects of temperature on rate constant has been adopted:

$$k_r = 6.32\times10^{-3}(1.066^{T-20} - 1.21^{T-60})\ f_m\ f_{fas}\ f_{ox} \qquad (4)'$$

where f_m = moisture factor, 0.85 at 50%

f_{fas} = free air space factor, 0.95 at 35%

f_{ox} = oxygen content in FAS factor, 0.82

From the above, the equation used to compute the rate constant for the proposed cell is,

$$k_r = 4.2\times10^{-3}(1.066^{T-20} - 1.21^{T-60})$$

The calculated values of the fractional biodegradation of mixed food as a function of temperature and time of reaction are shown in Figure 1.

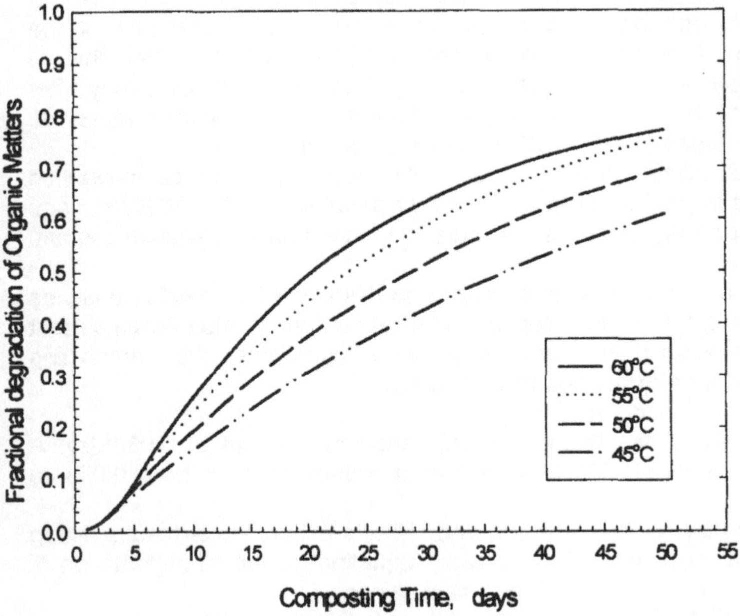

Figure1. Fractional bioconversion of OM with composting temperatures

Figure 1 shows how decomposition is expected to proceed with days of operation with time. The results show that, at 50°C, 28 days are required to finish the 50% degradation of organic matters, which is in fair agreement with the values reported by Sesay (1998), Muller (1998), and Murphy (1995). However, the initial very high values of decomposition reported by Sesay do not fit the first order reaction profile. Also, Figure 1 shows that increasing the composting temperature can shorten the required composting period. The projected relatively short bioconversion periods are in agreement with the reported results of aerated in-vessel systems, such as the agitated bins of Longwood Manufacturing Corporation (King, 1999), where the composting stage of biosolids mixed with leaves, paper fiber and wood shavings lasts 21 days, followed by a 24-day period of aerated curing of the compost product. Nearly the same composting and curing periods were reported for the treatment of biosolids mixed with wood chips and sawdust in aerated, static piles (Block, 1999) and are used in the aerobic composting facilities of Halifax, Nova Scotia (Wendt, 2000).

Heat generation and dissipation in an aerobic cell

The aerobic decomposition of organic waste, as represented by Equation (2) is a strongly exothermic reaction. The calculated heat of biodegradation is –614.6 kcal/mol (4,192 kcal/kg dry OM) which is nearly the same as that obtained from Equation 2 (Roinen, 1999). The air is assumed to enter the cell at 20°C and 50% humidity, and to leave at 50°C - 60°C and 100% humidity. Therefore, the net moisture removed with the exhaust gases is 0.079(50°C) – 0.153(60°C) g water/g inlet air. Furthermore, it is assumed that each ton of mixed food waste contains 640 kg of moisture and 360 kg of organic matter. The total heat generated by 50% biodegradation during the composting period is calculated to be 893 Gcal.

The corresponding daily heat generation pattern by biodegradation is shown in Figure 2. It can be seen that the maximum generation of heat occurs within 4-5 days after composting is started.

Kayhanian and Tchobanoglous (1992) reported that the typical bio-degradabilities of mixed food waste and mixed paper are 81.9% and 66.7%, respectively. For 50% bioconversion of the organic materials fed to the cell, 148 kg of biodegradable organic matters (BOM), i.e. 1.01kmol must be reacted per ton of feed material, over the composting period. In order for the cell to remain at steady state during the campaign, the amount of heat generated by reaction must be removed continuously. This can be accomplished by a) the upward air flow, b) the evaporation of the downward water flow into the ascending air stream, and c) the convection and radiation heat losses from the cell to the environment.

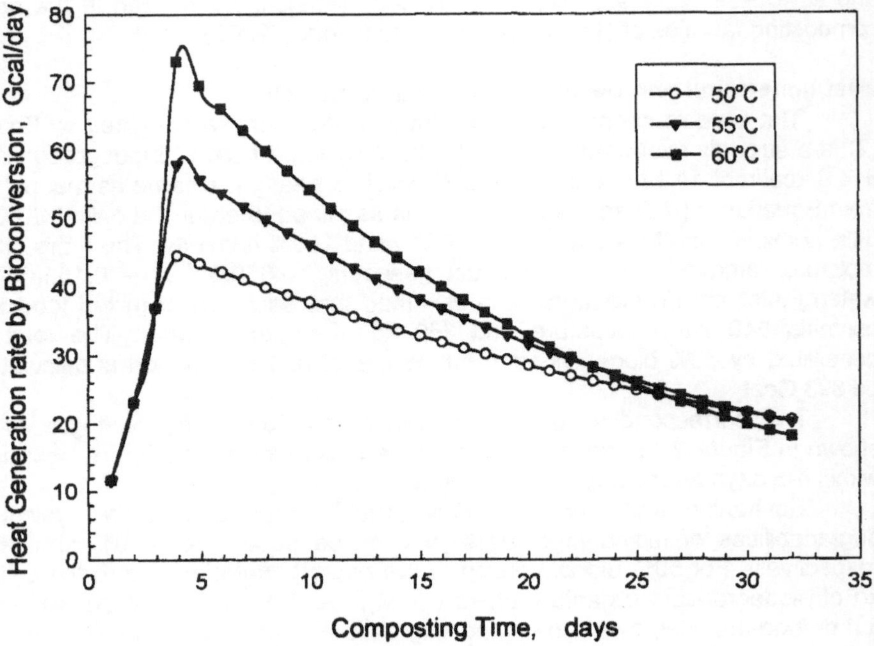

Figure 2. Heat Generation Rate of proposed ABC cell
(capacity 1,440 ton; 36%OM)

An approximate calculation showed that the principal heat loss is by natural convection/radiation from the surface of the bed and the outer surface of the walls of the cell (both assumed to be at 40°C) to the atmosphere (20°C). The overall heat transfer coefficient was estimated at 10 W/(m^2 K) (Themelis, 1995). The calculated overall heat loss during the entire composting period ranges from 227Gcal (60°C) to 252Gcal (50°C). This result indicates that only about one forth of the heat generated by bioconversion will be removed through heat loss to surroundings. Therefore,
it is necessary to cool the cell by means of air flow through it and the evaporation of water added to the bed during operation.

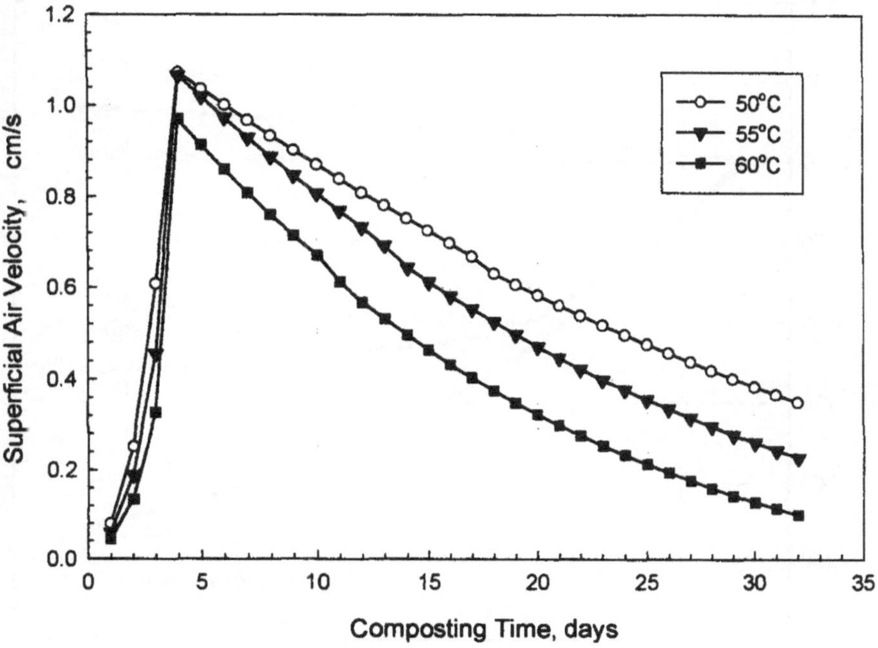

Figure 3. Aeration Requirement for proposed Cell
[Size:100mx6mx3m; Capacity:1,440 ton MFW(36% OM)]

The air flow through the cell was dictated by the requirements for thermal alance. Figure 3 shows how the required air supply through the bed, expressed s superficial velocity, varies during the composting period. As the operating emperature of bed increases, both average and peak air supply rates decrease rimarily due to the higher heat of vaporization. Within the operating emperatures considered, the average superficial velocity of air varies from .55cm/s (1,980cfh/ton of OM reacted; 60°C) to 0.67cm/s (2,400cfh/ton of OM eacted; 50°C) and the peak air velocities are 1.1 cm/s (3,950cfh/ton of OM eacted). The amounts of air supplied to keep the bed temperature 50°C and 0°C are calculated to be 55 and 30kg air/kg OM reacted, respectively, which is bout 9 – 5 times higher than the stoichiometric air requirements.

The excess air introduced to keep the temperature of the bed eventually auses drying of solid waste. To assure the proper rate of iodegradation reaction, moisture contents should be maintained over 50%. ontinuous spraying of water is therefore necessary at about one or two weeks fter composting has started, depending on the temperature of exit air and the

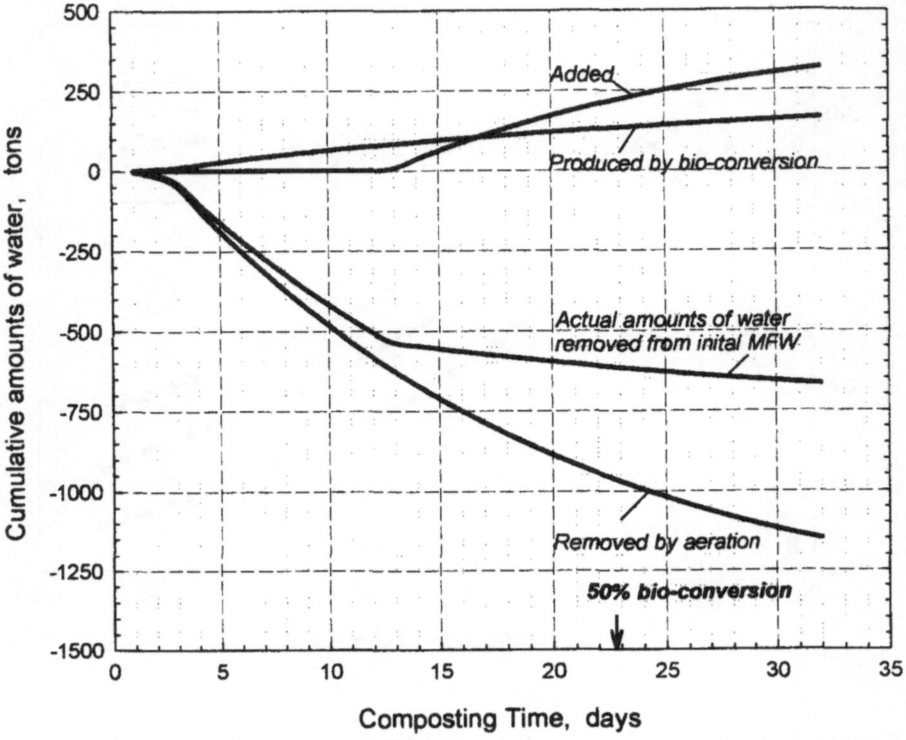

Figure 4. Water balance of ABC Cell [Temp. 55°C; Moisture level 50%; Capacity 1440 ton(36%OM)]

desired moisture level. The cumulative water balances at 55°C is presented in Figure 4.

Per each ton of organic matters, the initial amount of water is 1.78 tons and another 0.25 tons are produced by bioconversion and the amount of water removed from mixed food waste to keep the moisture contents 50% is 1.19 tons for all operation temperatures. On the other hand, the amount of water removed by aeration mainly relies upon the desired composting temperature. By increasing the composting temperature from 50°C to 60°C, the water addition required to keep the 50% moisture level throughout the composting period increases from 0.3 tons/ton of OM, to 2.5 tons/ton of OM. This result implies that the amount of water carried out by the saturated exit air is the most important factor to determine the amount of additional water required during the composting period.

Conclusions

This study used operating conditions recommended in a number of experimental studies to develop the process design of a 600 m^2 by 3-m deep aerobic bioconversion cell that can compost 1,440 metric tons of mixed food and wastes in 19-28 days period. The bioconversion plant envisaged would consist of a number of such cells, some in operation and some being loaded or unloaded. The re-use of the cells is predicated on the assumption that the compost product will be used in large-scale soil conditioning projects. It is assumed that the forced airflow through the bed will be sufficiently uniform to prevent the formation of anaerobic pockets where methane and noxious gases are produced.

The results of the study showed that, for a cell of this design, the heat losses to the surroundings over the composting periods would amount to 227 - 252Gcal, about 25% of the chemical heat released by aerobic bioconversion. The rest of this heat must be removed by means of an upward airflow through the cell that corresponds to 500 - 900% of the stoichiometric requirements for oxidation. The principal cooling is provided not by the sensible heat in the air but by the evaporation of the water percolating through the cell and saturation of the airflow. The average superficial velocities of the air are 0.67, 0.63 and 0.55 cm/s for the bed temperatures of 50°C, 55°C and 60°C, respectively. The conversion of these velocity to specific aeration rate, G, ranges from 2,000 – 2,400 cfh/ton OM, which is within the typical aeration rate of static pile systems.

This study has quantified the principal difference between anaerobic and aerobic biodegradation in a static cell, from a process design point of view. In anaerobic reaction, most of the chemical energy of the organic matter is transferred to the generated methane gas; this gas can be captured and then used as a fuel. In aerobic bioconversion in a static cell, this energy must be carried out in a continuous and relatively large air/water vapor flow through the cell. For the 600 m^2 x 3m cell of this study, the air to be injected for 50% bioconversion amounts to 4.5 - 8.0 tons/ton of feed materials, or 3.3–4.0 standard m^3/s for the entire cell. The air blower to deliver this flow at 1/3 atm (5 psig) would be rated at about 200-240kW. The corresponding power consumption over the entire period is calculated at 63kWh(60°C, 19days)-112kWh(50°C, 28 days) per ton of waste material to be treated; at present power rates this would amount to \$6-\$11/ton of mixed food waste and would represent the major operating cost. The water input required per ton of dry OM to maintaining the desirable moisture level in the bed and saturate the airflow would increases from 0.3tons to 2.5tons as the composting temperature increases from 50°C to 60°C.

The method of analysis presented above can be applied to any cell geometry.

REFERENCES

Bedminster Bioconversion Corp. (1995) Cherry Hill, NJ (Plants: Pinetop-Lakeside, AZ; Sevier County, TN).

Bernreuter, J. and Stessel, R.I. (1999) A Review of Aerobic Biocell Research and Technology, y, White Paper submitted to Biological Processes Subcommittee of SWANA (copy available, Earth and Environmental Engineering, Columbia University, New York, NY, 10027).

Block, D. (August 1999) Compost Plays Role in Riverfront Restoration, *Biocycle*, 26-29.

Dennison, R.A. (1996) Environmental life-cycle comparisons of recycling, landfilling, and incineration, *Annu. Rev. Energy Environ.*, Vol. 21, 191-237.

Franklin Associates (1995) The role of recycling in integrated waste management in the United States, Rep. *EPA/530-R-96-001*, USEPA, Munic. Industrial Waste Division, Wash. DC.

Graedel, T.E. and Allenby, B.R. (1995) *Industrial Ecology*, Prentice Hall, Englewood Cliff, NJ.

Green, L. (March 10, 1999) Method for treating biodegradable material through aerobic degradation, *U.S. Patent No. 5,882,022.*

Hamoda, M. F., H. A. Abu Qdais, et al (1998) Evaluation of Municipal Solid Waste Composting Kinetics, *Resources, Conservation and Recycling*, **23**: 209-223.

Haug, R.T. (1993) *The Practical Handbook of Composting Engineering*, chap.2, 10, Lewis Pub.

Hudgins, M. and Harper, S. (1998), Successful Demonstration of Two Aerobic Landfills- Leading Towards a More Sustainable Solid Waste Management Approach, *Environmental Control Systems, Inc.*, Aiken, SC.

Jewell, W.J. (August 1999) Latest Progress in Anaerobic Digestion, *Biocycle,* p. 64-66.

Kayhanian, M. and Tchobanoglous(1992) Computation of C/N ratios for various organic fraction, *BioCycle*, 33(5)

Keener, H.M., D.L. Elwell, et al.(1998) Specifying Design/Operation of Composting Systems Using Pilot Scale Data, *Applied Engineering in Agriculture*; Keener, H M., R.C. Hansen, et al. (1997) Airflow Through Compost: Design and Cost Implications, *Applied Engineering in Agriculture,* **13**(3): 377-384.

King, M.A. (August 1999), An Amazing 93% Reuse for Biosolids, *Biocycle*, 42-45

Markels, Jr. (Oct. 15, 1996), Method of improved landfill mining, *U.S. Patent No. 5,564,862.*

Merz, R.C. and Stone R. (January 1962) Sanitary Landfill Behavior in an Aerobic Environment, *Public Works* , 68-70.

Muller, W., K. Fricke, et al. (1998) Biodegradation of Organic Matter during Mechanical Biological Treatment of MSW, *Compost Science & Utilization*, **6**(3): 42-52.

Murphy, R. J., D. E. Jones, et al. (1995), ?Relationship of Microbial Mass and Activity in Biodegradation of Solid Waste, *Waste Management & Research*, **13**: 485-497.

Roinen, A. (1999) *HSC Chemistry 4.0*, Outokumpu Research, Outokumpu Oy, Pori, Finland.

Sesay, A.A., Lasaridi, K.E., and Stentiford, E.I. (1998) Aerated static pile composting of municipal solid waste, *Waste Management and Research*, 16:3, p. 264-272.

SCS Engineers (1992) *New York City Waste Composition Study (1989-1990)*, NYC Department of Sanitation, New York City, NY.

Tchobanoglous, G., Theisen, H., and Vigil, S. (1993) *Integrated Solid Waste Management*, Chapter 4, McGraw-Hill, New York.

Themelis, N.J. (1995) *Transport and Chemical Rate Phenomena,* Ch. 12, Gordon Breach Pub., New York.

Wendt, F. (Jan. 2000) private communication, Waste Resources, Halifax Regional Municipality, Halifax Nova Scotia, Canada.

COMPUTER MODELING OF WASTE INCINERATORS

Joseph Reynolds
Louis Theodore
Chemical Engineering Department
Manhattan College
Riverdale, NY 10471

INTRODUCTION

In 1988, in response to a request from EPA's Air Pollution Training Institute, a computer program to predict hazardous waste incinerator behavior was developed. The program, named *HWI*, employs stoichiometric, thermodynamic, and sizing calculations to determine operating temperature, flue gas composition and preliminary design information. The purpose of this paper is to describe recent modifications to this program.

The original *HWI* program has been described in the literature[1,2], and has received favorable reviews[3,4]. Since that time, it has been used for both training purposes (by EPA and academia) and for design/predictive calculations for existing and proposed units. In 1991, the program was made available to the public through Wiley-Interscience[5] and more recently (2001), a much expanded version, *HWIF*, has been published, again through Wiley-Interscience[6].

Since its initial development, *HWI* has undergone several major modifications. The first was in 1991 when, at the behest of a major hazardous waste incinerator company, the capability of the program was extended to include many options of particular interest to that company. This second version has also been described in the literature[7], but the modifications to the program are proprietary and are not publicly available. The second major modification was performed in 1997, again at the request of EPA's Air Pollution Training Institute. This new version is available again through Wiley-Interscience[6] as part of a new book on hazardous waste incineration. In 1999, the program was greatly expanded to include other equipment besides the incinerator. This latest version, known as *HWIF*, is also available[6] but is not covered in any detail in this paper. In the following sections, a brief review of the *HWI* program and two example problems with computer-generated solutions are presented.

THE *HWI* PROGRAM

As is the case with most computer programs, *HWI* is organized into three parts: input of data, calculations, and output of results. Each part is described in separate sections below.

Input

There are two different ways of inputting data to the *HWI*: (a) from the terminal and (b) through the use of an input data file.

(a) **Terminal input** is accomplished interactively; prompts appear on the monitor screen and the user answers them from the keyboard. However, if desired, the input data can be stored in an input data file for future use. This "data storage" takes place while the user is entering the data from the keyboard in answer to the prompts. The name and subdirectory path of the input data file is assigned by the user.

(b) The **input data file** generated during a previous run of *HWI* (as described above) can be used for data input, alleviating the necessity of retyping the input data at the keyboard. If changes to a particular data set are desired, any text (ASCII) editor may be used to alter the values in the input file.

The use of an input data file is ideal for "perturbation studies", where the effect of varying or perturbing certain operating parameters on incinerator performance is to be studied. Such a study is later demonstrated in Illustrative Example Problem 2. The user simply (1) enters the "base case" input data once by the usual method of answering the prompts while, at the same time, storing these data in an input file, (2) generates the "base case" results, (3) edits the "base case" input file to change the operating parameter to be perturbed, and (4) generates results for the "perturbed case". Steps (3) and (4) can be repeated as often as desired.

The data to be inputted by the user include: calculational mode, waste composition, waste flow rate, air requirements, waste heating value, heat loss from the incinerator, and some preliminary design information.

1. **Calculational mode**. The three key parameters in the *HWI* calculations are the incinerator operating temperature, the waste-fuel heating value and the amount of excess air. The waste-fuel heating value is determined by the materials being incinerated. Either one of the remaining two parameters may be treated as the independent variable. The *HWI* program provides two options: (1) the amount of excess air to be employed may be specified by the user, and the resulting incinerator operating temperature calculated by *HWI*; or (2) the desired operating temperature may be inputted by the user and the required amount of excess air calculated by the program.

2. **Waste composition**. In order to accommodate various methods of characterizing the chemical make-up of the waste, *HWI* provides the user with the following options: (a) elemental (or ultimate analysis), atom basis, (b) elemental analysis, mass basis, (c) componential analysis, mole basis, and (d) componential analysis, mass basis. The following elements are "recognized" or accepted by *HWI*: carbon, hydrogen, oxygen, nitrogen, sulfur, chlorine, fluorine, bromine and iodine.

Inert material (ash), which is assumed to exit the incinerator as solids (particulates) entrained in the flue gas stream, may also be inputted as part of the waste-fuel composition. This material is assumed to have negligible impact on the incinerator processes.

3. **Waste flow rate**. The waste (or waste-fuel mixture) flow rate is the next input requested. The user has the choice of using lb/hr or lbmol/hr.

4. **Air requirements**. If Calculational Mode 1 is being used (i.e., the user inputs the amount of excess air and *HWI* calculates the operating temperature), the amount of excess air is inputted at this point. If Calculational Mode 2 is being used (i.e., the user inputs the desired operating temperature and *HWI* calculates the required amount of excess air), only the temperature of the preheated air is called for.

5. **Waste heating value**. The net heating value (NHV) of the waste (also referred to as the lower heating value, and defined as the enthalpy change of stoichiometric combustion when the reactants and flue products are at the same reference temperature, usually 60°F, and any water present in the product stream is in the vapor state) may either be inputted by the user or estimated by Dulong's method [6].

6. **Heat loss**. In practice, any heat loss from the incinerator is often neglected. If the heat loss is to be taken into account, a prompt is provided for its input in terms of percent of the NHV.

7. **Preliminary design data**. The last set of input data is required for a preliminary design (sizing) of the incinerator. For this purpose, either the superficial throughput velocity of the flue gas or the length-to-diameter ratio (L/D) of the incinerator must be specified by the user. If the user omits this input, a L/D of 3.0 is assumed. The heat release rate, which is a function of the type of incinerator and, to a lesser extent, the nature of the waste, is another required design parameter. If the user ignores this input, a release rate of 25,000 Btu/hr-ft^3 (a fairly typical value for many types of incinerators) is assumed.

Calculations

After the last input, the program performs stoichiometric and thermodynamic calculations.

Stoichiometric Calculations

First, an initial determination of the flue gas composition is performed. The flue gas consists of the products of combustion of the waste plus nitrogen and any excess oxygen from the air. With the exception of chlorine and sulfur, all combustion reactions are assumed to go to completion. The combustion products for each of the elements in the waste listed earlier are assumed to be CO_2, H_2O, HCl, Cl_2, O_2

(excess), N_2, SO_2, SO_3, HF, Br_2 and I_2. Note that carbon is completely oxidized to carbon dioxide and nitrogen is not oxidized at all (although there are methods for estimating the oxides of nitrogen[8]). Hydrogen chloride and sulfur dioxide are not completely stable and the program takes into account the chemical equilibrium between HCl and Cl_2 as well that between SO_2 and SO_3. This is explained in more detail later.

Enthalpy Calculations

In order to calculate the operating temperature, the conservation laws for mass and energy must be employed in conjunction with thermochemical principles. A complete enthalpy balance may be applied around the incinerator accounting for all temperature changes and latent (phase) and combustion (reaction) enthalpy effects. Since heat (enthalpy) of reaction and/or net (or gross) heating values are generally available at a reference temperature (T_o), the calculations proceed along a chosen path that includes the reaction at this reference temperature. The total enthalpy change is therefore considered to have the following contributions: (1) the cooling of the feed to T_o, (2) the combustion reaction at T_o, (3) the heating of the flue products to the incinerator temperature, T, and (4) the heat loss from the incinerator.

The enthalpy balance can be used to provide a rigorous estimate of the incinerator operating temperature, T. However, the contribution of (3) above cannot be determined exactly until the flue gas composition is more accurately known. (Up to this point in the calculations, Cl_2 and SO_3 formations have been ignored.) Using the approximate flue gas composition obtained by ignoring these equilibria, a initial estimate of the temperature may be made.

Equilibrium Calculations

The calculation of the flue gas composition by the stoichiometric method shown earlier is not difficult. However, the hydrogen chloride and sulfur dioxide are involved in equilibrium reactions, as indicated above. These reactions are:

$$2\,HCl + \tfrac{1}{2}\,O_2 \rightleftharpoons Cl_2 + H_2O \qquad (1)$$

$$SO_2 + \tfrac{1}{2}\,O_2 \rightleftharpoons SO_3 \qquad (2)$$

It should be noted that any of the other halogens present in the waste will, like chlorine, produce combustion products that are involved in equilibrium reactions. At normal incinerator temperatures, however, these equilibria heavily favor HF over F_2, Br_2 over HBr, and I_2 over HI. As a result, the concentrations of F_2, HBr, and HI in the flue gas are considered negligible.

The equilibria of Eqns. (1) and (2) make the calculations much more complicated because the equilibrium concentrations of chlorine and sulfur trioxide depend on the incinerator temperature. The temperature achieved in the incinerator, however, depends to some extent on the flue gas composition, because almost all of the energy released by the incineration reactions is absorbed by the flue gas, the heat capacity of which depends on its composition.

These equilibria effects are often neglected in calculations to predict the flue gas composition because the amounts of sulfur trioxide and chlorine formed are generally small and the equilibrium calculations are cumbersome. However, these two products, in even small amounts, can significantly impact on the design and performance of an incinerator facility and the extent of their presence in the flue gas should be determined.

For gas phase reactions, the equilibrium constant, K, may be approximately represented in terms of the partial pressures of the components involved. For the reactions provided in Eqns. (1) and (2), these relationships are given by Eqns. (3) and (4), respectively.

$$K_1 \approx K_{p,1} = \frac{p(Cl_2)\,p(H_2O)}{p(HCl)^2\,p(O_2)^{1/2}} \quad (3)$$

and

$$K_2 \approx K_{p,2} = \frac{p(SO_3)}{p(SO_2)\,p(O_2)^{1/2}} \quad (4)$$

where $p(i)$ = partial pressure of component i
K_p = approximate equilibrium constant based on partial pressures

The calculations become more complicated when both chlorine and sulfur are present in the waste. Since the partial pressure of oxygen appears in both Eqns. (3) and (4), these equations must be solved simultaneously.

From standard heat of formation data, Theodore and Reynolds [6] have obtained the chemical reaction equilibrium constants K_1 and K_2 as functions of temperature. Using these relationships, the two equilibrium constants are determined for the temperature estimated earlier. Numerical methods are then used to solve Eqns. (3) and (4) simultaneously and the flue gas composition is recalculated, this time including the sulfur trioxide and chlorine. A more accurate incinerator temperature is next calculated from the enthalpy balance, but this time, the enthalpy change for heating the products is evaluated using the corrected flue gas composition. (Note that the heat capacities of the components of the flue gas are functions of temperature, and this calculation involves an integration of the heat capacities from room temperature to the incinerator temperature.) New equilibrium constants and a new flue gas composition based on this revised temperature are next calculated. If necessary, this process or loop is repeated several times until the calculated temperature, equilibrium constants, and compositions converge to constant values.

Preliminary Design Calculations

The incinerator temperature and flue gas composition are used to determine the flue gas volumetric flow at actual (acfm), standard (scfm) and dry standard (dscfm) conditions, as well as the molar flow rate. An estimate of the incinerator volume is obtained from the volumetric flow rate and the heat release rate. Assuming the incinerator to be cylindrical, the dimensions are calculated from either the L/D ratio or the superficial velocity (whichever one was inputted by the user). The required residence time is also calculated.

Output

Output from the *HWI* program, which consists mainly of tables of calculated results, is written to an output data file, which may later be sent to a printer, and also appears on the monitor screen. Two of those tables are shown in Figures 1 and 2. The first gives the composition of the flue gas leaving the incinerator and the second gives information such as flow rates, temperature, excess air, SO_3 and Cl_2 concentrations in ppm (parts per million by volume), and HCl mass flow rate. If ash is present, the particulate loading in gr/acf (grains per actual cubic foot) and gr/dscf (grains per dry standard cubic foot) is also given. The latter is corrected to 50% excess air (approximately 7% excess oxygen), that is, the tabulated value is the particulate loading in gr/dscf that would be carried by the flue gas if 50% excess had been used.

FLUE GAS COMPOSITION

COMPONENT	LB*	MASS %	LBMOL*	MOLE %
CARBON DIOXIDE	122.01	11.69	2.772	7.91
WATER (g)	33.41	3.20	1.854	5.29
NITROGEN	728.01	69.73	25.991	74.15
OXYGEN	95.92	9.19	2.998	8.55
HYDROGEN CHLORIDE	36.13	3.46	.991	2.83
CHLORINE	.12	.01	.002	.00
SULFUR DIOXIDE	28.40	2.72	.443	1.26
SULFUR TRIOXIDE	.10	.01	.001	.00
TOTAL	1044.10	100.00	35.052	100.00

* Basis: 100 lb waste (or waste-fuel mixture).
Note: Contribution of particulates (fly ash) not included.

Figure 1. Output file for Example 1: Flue Gas Analysis

FLUE GAS DATA	
Mass flow rate, lb/hr	34620.
Volumetric flow rate, acfm	48708.
Volumetric flow rate, scfm (60 F)	7320.
Volumetric flow rate, dscfm (60 F)	6933.
Molar flow rate, lbmol/hr	1157.
Temperature, F	3000.
Percent excess air	76.7
Sulfur trioxide concentration, ppm	3.512E+01
Chlorine concentration, ppm	4.956E+01
Hydrogen chloride concentration, ppm	2.827E+04
HCl mass flow rate, lb/hr	1192.2
Particulate loading, gr/acf	.40
Particulate loading, gr/dscf -- corrected to 50% excess air	3.27

Note: Molar flow rate does not include particulate matter.

Figure 2. Output file for Example 1: Flue Gas Data

ILLUSTRATIVE PROBLEMS

To demonstrate the use of *HWI*, two example problems are presented in this section.

Example Problem 1

The waste described below is to be incinerated at 3000°F at a rate of 3,300 lb/hr using air preheated to 300°F. Calculate (a) the amount of excess air to be used, (b) the composition of the flue gas, and (c) the size of the incinerator (length, diameter) for a L/D ratio of 2.0. Assume a heat release rate of 25,000 Btu/hr·ft^3.

constituent	mass %	NHV (cal/gmol)
ethylene chloride	49.2	670,500
thiophene (C_4H_4S)	37.4	271,000
water	8.4	-10,519
ash	5.0	0

Solution

Part of the output for this problem is shown in Figures 1, and 2. The answers to questions (a) and (b) can be found in these tables. The answers are: (a) 76.7% excess air should be used; (b) the flue gas consists of 11.69% (mass) CO_2, 3.20% H_2O, 69.73% N_2, 9.19% O_2, 3.46% HCl, 0.01% Cl_2, 2.72% SO_2, and 0.01% SO_3; and (c) the incinerator is 17.6 feet in length and 8.8 feet in diameter. The output table showing the Preliminary Design results is not shown.

Example Problem 2

For the waste described in Example Problem 1, determine the effect of changing the water content of the waste on (a) the incinerator temperature, and (b) the chlorine and sulfur trioxide concentrations of the flue gas. Use the amount of excess air found in answer to part (a) of Example Problem 1. As the water content changes, the ethylene chloride, thiophene and ash concentrations are to remain in the same relative proportions.

Solution

This is an example of a "perturbation study" mentioned earlier. To generate the results, an input data file containing information similar to that of Example Problem 1 was first generated. The only difference between this input file and that of Problem 1 was that Calculational Option 1 (the amount of excess air is known and the temperature is calculated) was used. The excess air was set at 76.7%, which was found in the output of Problem 1. That input file was then used a number of times, each time changing only the mass percent of water. (It was not necessary to adjust the mass percents of the ethylene chloride, thiophene and ash because the program automatically "normalizes" the input composition, i.e., if the numbers input by the user do not add up to either 100 or 1, they are scaled up or down to do so.) After each run the chlorine and sulfur trioxide contents of the flue gas as well as the incinerator temperature were recorded. These results are shown in Figures 3 and 4. In Figure 3, the temperature decreases with increasing water content for the obvious reason that the NHV of the waste is decreasing. Besides decreasing the concentrations of the combustible components (ethylene chloride and thiophene), the water *absorbs* heat when it vaporizes, making less available to heat the products to the final temperature.

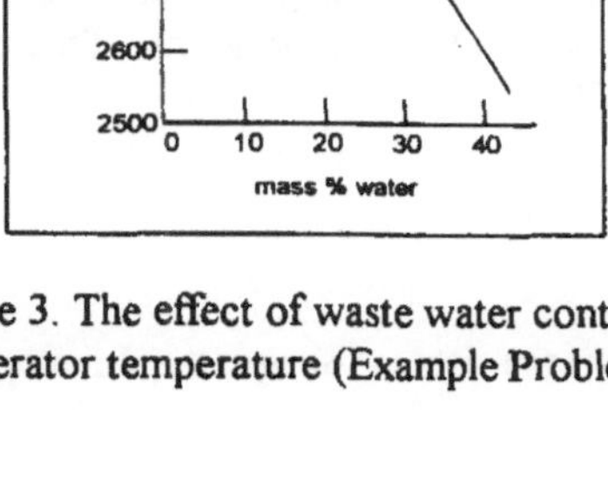

Figure 3. The effect of waste water content on incinerator temperature (Example Problem 2).

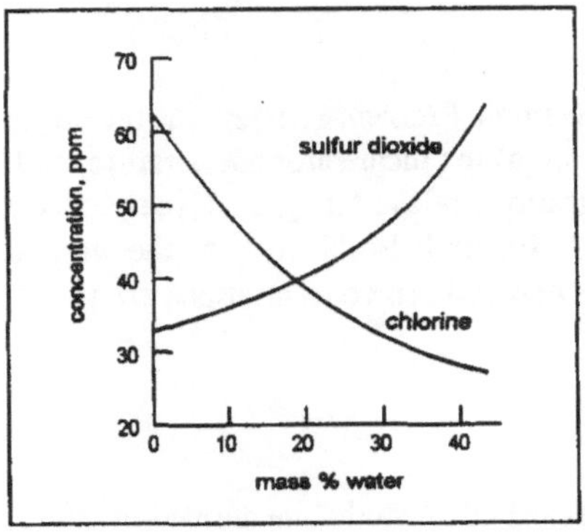

Figure 4. The effect of waste water content on the chlorine and sulfur trioxide concentrations of the flue gas (Example Problem 2).

In Figure 4, the sulfur trioxide concentration increases with water content as a direct result of the decrease of temperature. The oxidation of sulfur dioxide is exothermic, which means that the equilibrium constant, K_2, increases with decreasing temperature; this shifts the equilibrium toward the sulfur trioxide, as shown by Eqn. (4). Figure 4 also shows that increasing water content decreases the chlorine concentration of the flue gas. The oxidation of HCl is also exothermic, which fact alone would encourage chlorine production. It should be obvious from Eqn. (3), however, that since water is a product of the reaction, an increase of its partial pressure shifts the equilibrium away from the Cl_2. The second effect outweighs the first, and the net effect is a drop in the flue gas chlorine concentration.

THE EXPANDED PROGRAM, HWIF

This expanded version of the model includes many new features, among which which are: a secondary combustion chamber, a quencher/tempering chamber, N_2-NO and CO_2-CO equilibria, oxygen-enriched air, humidity calculations, and a third calculational option to determine supplemental fuel requirements. The last feature allows the user to input the amount of excess air and the desired incinerator temperature; the program then calculates the amount of supplemental fuel to be used. The options for type of fuel are natural gas, #2 fuel oil, or a user-defined fuel the composition of which is an input to the program. This program is available through Wiley-Interscience[6].

ACKNOWLEDGMENT

This work was performed at Manhattan College; the development of the original program as well as the modifications reported on in this paper were supported by grants from the Office of Air Quality Planning and Standards of the U.S.E.P.A.

REFERENCES

1. J. Reynolds and L. Theodore, "A Computerized Simulator for Hazardous Waste Incineration", Paper No. 88-51.3, *Proceedings of the 81th Annual APCA Meeting*, Dallas, 1988.

2. J. Reynolds, L. Theodore, "Hazardous Waste Incinerator Calculations", *Proceedings of the 1988 International Specialty Conference on Environmental Challenges in Energy Utilization*, Cleveland, 1988.

3. M. Ruby, "Microcomputer Software Reviews", *Journal of the Air and Waste Management Association*, June, 1989.

4. A. Kumar and H. S. Huang, "HWI: A Software for Hazardous Waste Incinerator Calculations", *Environmental Progress*, November, 1989.

5. J. Reynolds, R. DuPont, L. Theodore, *Hazardous Waste Incineration Calculations: Problems and Software*, Wiley-Interscience, New York, NY, 1991.

6. J. Santoleri, J. Reynolds, L. Theodore, *Introduction to Hazardous Waste Incineration*, 2nd ed., Wiley-Interscience, New York, 2001.

7. J. Reynolds, P. Farber, C. Pratt, L. Theodore, "Computer Calculations for a Hazardous Waste Incineration Facility", Paper 91-31.10. *Proceedings of the 84th Annual AWMA Meeting*, Vancouver, 1991.

8. L. Theodore, E. Moy, A Theodore Tutorial, *Hazardous Waste Incineration*, ETS International, Roanoke, VA, 1995.

SYMPOSIUM 5:

Water Quality Modeling

IMPACT OF HIGHWAY DEICING AGENT LOADING ON WATER QUALITY

DAVID OSTENDORF, JENNIFER MACKEY, and ERICH HINLEIN
Civil and Environmental Engineering Department
University of Massachusetts
Amherst, MA 01003

INTRODUCTION

We monitored and model precipitation, runoff, and runoff chloride concentrations from the outdoor Loading Area in front of a covered Salt Storage Facility in Eastern Massachusetts. The environmental impacts of runoff from highway deicing applications [1-3] and benefits of indoor deicing agent storage are long recognized and well documented [4-6]. Our study quantifies the impact of spillage from uncovered delivery and loading on surface water quality, impacts that would be mitigated by a larger structure placing these operations indoors.

SITE DESCRIPTION AND METHODS

We analyze hyetographs, hydrographs, and pollutographs from the Loading Area of a Salt Storage Facility in Eastern Massachusetts, operated by the Massachusetts Highway Department (Figure 1). Salt (sodium chloride) was delivered to the Facility throughout the 2000-2001 deicing season by 15 to 25 ton capacity trucks, which dumped their contents onto the Loading Area, comprised of weathered asphalt pavement of 1,500 m^2 area A. A front end loader with a 3 ton bucket capacity then transferred the salt from the Loading Area to two covered Storage Sheds. Thus a residual layer of deicing material remained on the Loading Area from each delivery. The loader deposited salt from the Sheds into 10 to 15 ton spreader trucks during deicing applications. Any spillage during this loading added salt to the Loading Area as well.

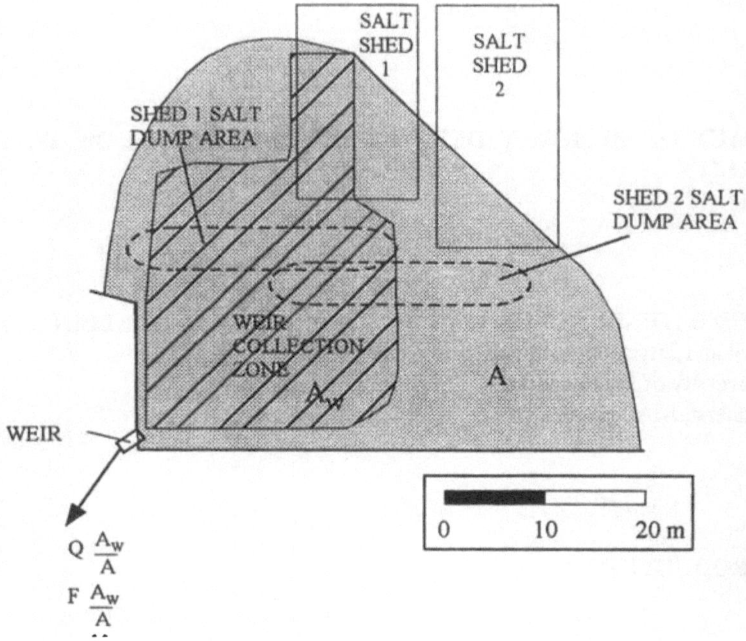

FIGURE 1. SALT STORAGE FACILITY LOADING AREA

A 90° V notch weir collected surface runoff from the Loading Area. The weir was equipped with an ISCO 4230 flow meter and a YSI 600 water quality probe that measured head and specific conductivity. The head determines water discharge through a classical weir equation [7], calibrated with volumetric data observed episodically over the period of record. We analyzed monthly surface water samples from the weir by ion chromatograph in order to calibrate a regression between the specific conductivity and the dissolved chloride concentration [8]. Precipitation was observed in an ISCO 674L tipping bucket rain gauge, equipped with heating mats to melt frozen precipitation. Weir and precipitation data were observed on 5 and 6 minute intervals respectively, with data sent over dedicated phone lines to an acquisition system in the laboratory. A smoothing subroutine distributes the precipitation P_I over 5 minute time intervals t_I synchronous with the discharge Q_I and chloride concentration C_I.

CHLORIDE DISSOLUTION MODEL

Figure 2 displays our lumped parameter model of water and chloride transport through the Loading Area. The conservation of solid chloride mass S (kg/m^2) is a balance of storage and dissolution during storms

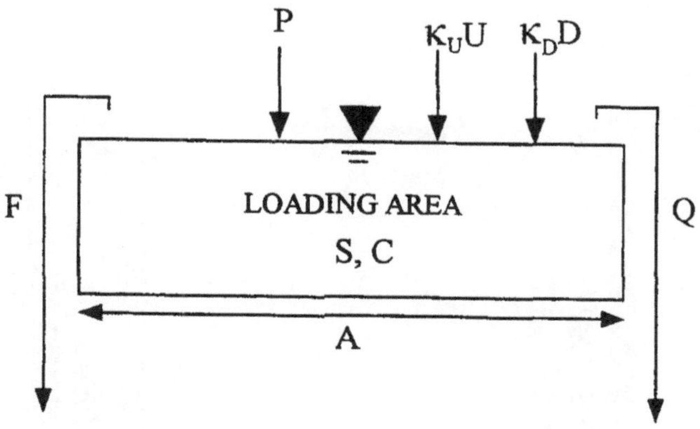

FIGURE 2. LUMPED PARAMETER LOADING AREA MODEL

$$A\frac{dS}{dt} + A\Omega S = 0 \tag{1a}$$

$$\frac{dS}{dt} = \frac{dS}{d\xi}P \tag{1b}$$

$$S = c_1 \exp(-\frac{\Omega\xi}{P}) \tag{1c}$$

with integration constant c_1 and first order dissolution rate Ω. We assume a constant rainfall rate P (m/s) and so change independent variables from storm time to accumulated precipitation depth ξ. The latter variable is appropriate because dissolution occurs only when there is water on the Loading Area. Equation (1c) suggests that the present (depth ξ, time t) concentration S_J of solid chloride spilled from an application U_J or delivery D_J at prior depth ξ_J (time t_J) is given by Equation (2a)

$$S_J = (\frac{\kappa_U U_J + \kappa_D D_J}{A})\exp[-\frac{\Omega(\xi - \xi_J)}{P}] \tag{2a}$$

$$S_I = \sum_{J=0}^{I}(\frac{\kappa_U U_J + \kappa_D D_J}{A})\exp[-\frac{\Omega(\xi_I - \xi_J)}{P}] \tag{2b}$$

with application κ_U and delivery κ_D spillage coefficients. We sum all prior applications and deliveries and recover the present (I subscript) solid chloride

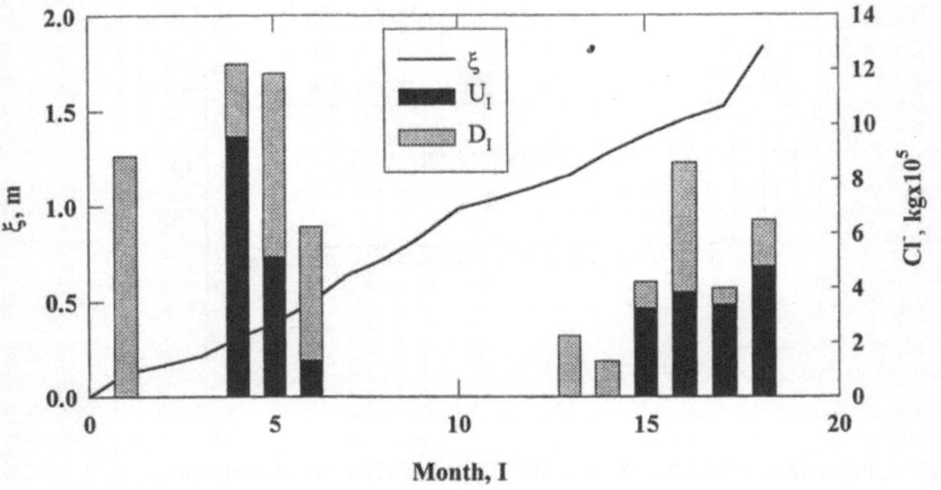

FIGURE 3. PRECIPITATION DEPTH ξ, MONTHLY APPLICATIONS U_I, AND MONTHLY DELIVERIES D_I AT SALT STORAGE FACILITY

concentration as a convolution integral, approximated on a monthly basis by Equation (2b). Figure 3 summarizes applications, deliveries, and precipitation since the onset of the 1999-2000 salting season. October 1999 is the first month of record (I equals 1).

We ignore the storage term in the conservation of water mass mass through the Loading Area, resulting in a balance of discharge and precipitation

$$Q = PA \tag{3}$$

The corresponding dissolved chloride balance [Equation (4a)] equates advective chloride efflux F and a dissolving source term

$$F = A\Omega S \tag{4a}$$

$$F = QC \tag{4b}$$

We combine Equations (3) and (4) on a monthly basis, then solve for the predicted present concentration C_I [Equation (5a)]. Thus, the dissolved chloride

TABLE I. OBSERVED MONTHLY WATER AND SALT BUDGET 2000-2001 DEICING SEASON

Month (I)	C_I (kg/m^3)	ξ_I (m)	F_I (kg)	U_I (10^5 kg)	D_I (10^5 kg)
April 00 (7)	1.13	0.641	85	0	4.87
May 00 (8)	0.234	0.727	23	0	0
June 00 (9)	0.0395	0.841	21	0	0
July 00 (10)	0.0651	0.987	26	0	0
August 00 (11)	0.0241	1.041	4.6	0	0
September 00 (12)	0.0240	1.099	4.6	0	0
October 00 (13)	0.389	1.167	125	0	1.30
November 00 (14)	0.532	1.277	102	0	0.942
December 00 (15)	5.88	1.372	1,260	3.29	0.358
January 01 (16)	11.6	1.453	1,270	3.88	4.73
February 01 (17)	10.2	1.520	751	3.40	0.584
March 01 (18)	2.80	1.834	1,700	4.82	1.65

concentration is the accumulated result of past solid chloride spillage on the Loading Area [Equation (2b)]

$$C_I = \frac{\Omega S_I}{P} \tag{5a}$$

$$C_I = \frac{\Omega}{AP}\sum_{J=0}^{I}(\kappa_U U_J + \kappa_D D_J)\exp[-\frac{\Omega(\xi_I - \xi_J)}{P}] \tag{5b}$$

MODEL CALIBRATION

Table I lists observed monthly Loading Area C_I and F_I values inferred from logged weir observations from April 2000 through March 2001. Observed discharges and chloride fluxes are scaled up by the ratio A to the weir collection area A_W of 650 m^2 (Figure 1), then integrated for each month. The resulting F_I values are listed in the Table. We note that the total annual observed chloride flux of 5,370 kg is 0.18% of the total applied ($1.55x10^6$ kg) and delivered ($1.44x10^6$) mass of chloride for the 2000-2001 deicing season. The flux responded to an annual precipitation depth of 1.14 m.

Table I also lists the observed C_I values implied by the hydrographs, pollutographs, and Equation (4). These measurements calibrate the solid chloride driven model [Equation (5b)], with the results summarized in Figure 4. We use a nested Fibonacci search [9] through these data to calibrate values for dissolution rate, application coefficient, and delivery coefficient

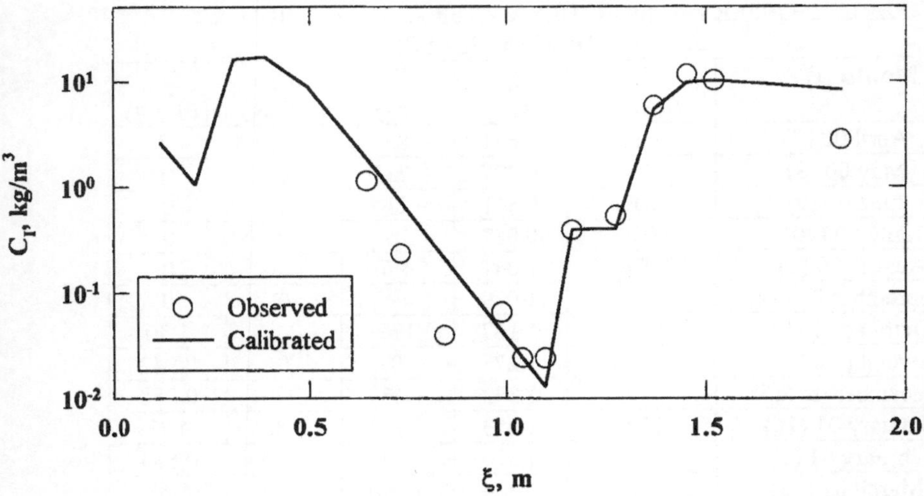

FIGURE 4. OBSERVED AND CALIBRATED DISSOLVED MONTHLY AVERAGE CHLORIDE CONCENTRATIONS

$$\Omega = 4.32\text{x}10^{-6}\,\text{s}^{-1} \tag{6a}$$

$$\kappa_U = 0.00216 \tag{6b}$$

$$\kappa_D = 0.000410 \tag{6c}$$

The optimal parameters minimize the mean error δ_M defined by

$$\delta_M = \frac{2}{N}\sum_N \left| \frac{C_I(\text{measured}) - C_I(\text{predicted})}{C_I(\text{measured}) + C_I(\text{predicted})} \right| \tag{7}$$

with N sampling points. The mean error of 45% is reasonable in view of the simplicity of the theory and the variation of C_I over three orders of magnitude. In the latter regard, both calibrated and observed concentrations rise abruptly from levels below 0.1 kg/m^3 to values in excess of 10 kg/m^3.with the onset of the deicing season in October (ξ_{13}=1.2 m). The slow dissolution implies that chloride concentrations persist throughout the year on the Loading Area, though at substantially reduced levels.

The calibrated dissolution rate rests on an average precipitation intensity of 4x10^{-7} m/s, inferred from the hyetographs recorded on the site [10]. It is close

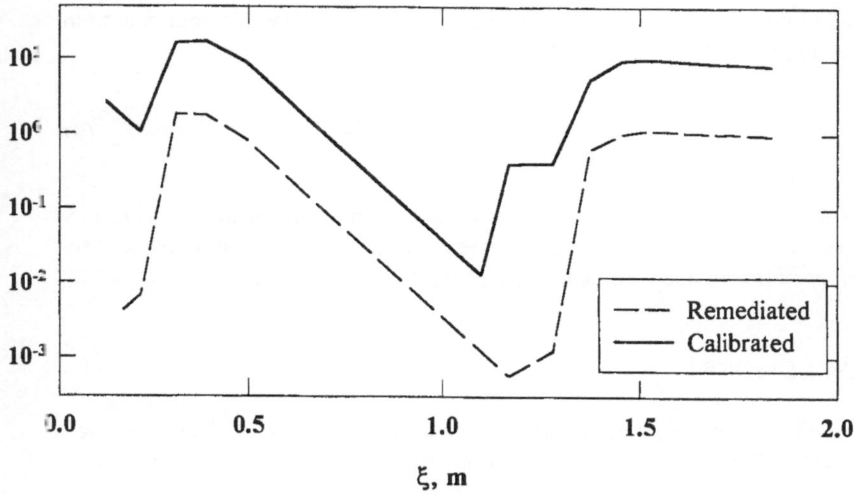

Note: Based on precipitation, deliveries, and applications of Figure 3.

FIGURE 5. CHLORIDE CONCENTRATIONS FROM OUTDOOR (CALIBRATED) AND INDOOR (REMEDIATED) LOADING AREAS

to the $2x10^6$ s^{-1} value found by Ostendorf et al. [10], who analyzed hyetographs, hydrographs, and pollutographs from the closed drainage system adjacent to the study area. The drainage system serves the Access Road but receives no spillage contaminated runoff from the Loading Area. Rather, Access Road chloride persists from deicing applications only.

DISCUSSION

The sum of the two calibrated spillage coefficients (0.25%) approximates the ratio (0.18%) of the annual chloride flux to the delivery and applied fluxes. We compare the calibrated κ_D with the ratio of $A^{1/2}$ to the 90 lane miles L served by the Facility

$$\frac{\sqrt{A}}{L} = 0.00027 \tag{8}$$

Equation (8) estimates the chloride applied to deice the Loading Area during storms as an order of magnitude less than the calibrated κ_U of 0.00216. This suggests that spillage associated with loading operations far exceeds the salt spread rate at the Salt Storage Facility.

Figure 5 assesses the reduction in chloride efflux realized by eliminating spillage by indoor delivery and loading. We simply replace the source term in Equation (5b) with the application rate implied by Equation (8)

$$C_I = \frac{\Omega}{L\sqrt{AP}} \sum_{J=0}^{I} U_J \exp[-\frac{\Omega(\xi_I - \xi_J)}{P}] \qquad (9)$$

Thus covered deliveries and loadings induce no contamination. Figure 5 suggests that the maximum Loading Area concentrations would be reduced by an order of magnitude if spillage were eliminated at the Facility.

CONCLUSIONS

Precipitation, runoff, and runoff chloride data suggest that spillage associated with salt delivery and loading operations far exceeds application rates on an outdoor Loading Area in front of a covered Salt Storage Facility in Eastern Massachusetts. About 0.2% of the delivered and applied chloride for the 2000-2001 deicing season ran off the Loading Area. A calibrated chloride dissolution model yields a solid chloride dissolution rate of $4.3\text{x}10^{-6}\ s^{-1}$. These slow kinetics ensure that spillage remains on the Loading Area throughout the year, though at substantially reduced levels during the summer months. A model simulation indicates that covering the Loading Area will reduce chloride runoff by an order of magnitude at the Facility.

ACKNOWLEDGMENTS

The Massachusetts Highway Department funded this research as part of Interagency Service Agreement 7721 with the University of Massachusetts at Amherst. The views and findings contained in this paper are the authors, and do not necessarily reflect official policy of the Massachusetts Highway Department. This paper does not constitute a standard, specification, or regulation.

REFERENCES

1. Gelhar, L.W. and J.L. Wilson. 1974. "Groundwater Quality Modeling," *Groundwater*, 12(6): 399-408.

2. Jones, P.H. and B.A. Jeffrey. 1992. "Environmental Impact of Road Salting," in *Chemical Deicers and the Environment*, D'Itri ed., Chelsea, MI, Lewis Publishers, pp.1-116.

3. Granato, G.E. 1996. "Deicing Chemicals as Source of Constituents of Highway Runoff," *Trans. Res. Rec.* 1533: 50-58.

4. Richardson, D.L., C.P. Campbell, R.J. Carroll, D.I. Hellstrom, J.B. Metzger, P.J. O'Brien, and R.C. Terry. 1974. "Manual for Deicing Chemical Storage and Handling," *EPA 670/2-74-033*. Washington, DC: US Environmental Protection Agency.

5. Transportation Research Board. 1991. "Highway Deicing: Comparing Salt and Calcium Magnesium Acetate," *Special Rept No 235*. Washington, DC: National Research Council.

6. Transportation Research Board. 1999. "Best Management Practices for Environmental Issues Related to Highway and Street Maintenance," *NCHRP Synthesis No 272.* Washington DC: National Research Council.

7. Potter, M.C. and D.C. Wiggert. 1997. *Mechanics of Fluids.* Upper Saddle River, NJ: Prentice-Hall, p. 475.

8. Granato, G.E. and K.P. Smith. 1999. "Estimating Concentrations of Road Salt Constituents in Highway Runoff from Measurements of Specific Conductance," *WRIR 99-4077.* Northboro, MA: US Geological Survey.

9. Beveridge, G.S.G. and R.S. Schechter. 1970. *Optimization: Theory and Practice.* New York NY: McGraw-Hill, p. 190.

10. Ostendorf, D.W., D.C. Peeling, T.J. Mitchell, and S.J. Pollock. 2001. "Chloride Persistence in a Deiced Access Road Drainage System," *J. Env. Qual.*, in press.

APPLICATION OF THREE DIMENSIONAL HYDRAULIC AND WATER QUALITY MODEL IN THE ASSESSMENT OF ENVIRONMENTAL IMPACTS RELATED TO THE PROPOSED DEEPENING OF NEW YORK AND NEW JERSEY HARBOR NAVIGATION CHANNELS

Robert M. Aiello, Santiago Alfageme, Guy A. Apicella, John Headland
New York-New Jersey Harbor Partnership
One Blue Hill Plaza, Box 1509
Pearl River, New York 10965-8509

Mark F. Lulka
U.S. Army Corps of Engineers-New York District
CENAN-PL-EA
26 Federal Plaza
New York, New York 10278-0090

ABSTRACT: A three-dimensional (3D) hydrodynamic and water quality model was applied to the New York and New Jersey Harbor as part of the NY-NJ Harbor Navigation Study (HNS). The HNS - a comprehensive examination of the need for navigation improvements within the Port of New York and New Jersey - evaluated navigation channel improvements including the deepening of existing Federal and non-Federal channels and channel designs that would facilitate an economically efficient and environmentally sound port for current and future requirements. The purpose of the 3D-hydrodymanic and water quality modeling effort was to examine the impacts of channel improvements/deepening on physical and chemical parameters that directly influence the biological community. The parameters were: tidal current and elevation modifications, water temperature, dissolved oxygen, and salinity. The hydraulic and water quality model results were used in an assessment of project impacts on the aquatic community through a modified guild approach.

The water quality modeling was performed using MIKE3, a general 3D modeling system developed by the Danish Hydraulic Institute (DHI). The model boundaries extend to approximately Atlantic City, New Jersey, to the south, Shinnecock Inlet and Orient Point (on the Atlantic Ocean and Long Island Sound, respectively) to the east, and the Hudson River estuary to the Federal Dam at Troy, New York. A nested grid system was used in order to facilitate refinement of specific areas of interest - horizontal resolution could be varied from 75 to 2025 meters.

Following model calibration and verification, the model was used to simulate three channel-deepening conditions: 1) existing conditions; 2) baseline, or future without-project conditions (reflects navigation channels at currently authorized depths); and 3) future with-project conditions (channel depths of 50-53 feet). Two summertime freshwater input flow conditions were examined for each channel condition, low freshwater input (low-flow [min30Q10]) and high freshwater input (high-flow [max30Q10]). Impacts of the proposed deepening project are associated with the difference between future and baseline conditions.

The proposed future with-project conditions will produce minor changes in water surface elevations under low-flow conditions (i.e., maximum difference in tidal range is less than 1.6 inches). The changes in salinity between with- and without project conditions for the low-flow condition are small with a maximum average change of 0.7 ppt, which occurred in the Arthur Kill near Howland Hook. The maximum average salinity change in the remaining study area is 0.2 ppt or less. Overall, the average differences in dissolved oxygen between future and baseline conditions are very small, with a maximum reduction of 0.13 mg/l. Additionally, it should be noted that dissolved oxygen increases are predicted for some areas of the Harbor. The predicted changes in salinity and dissolved oxygen are very small in comparison to normal seasonal and annual variations in Harbor conditions - they are not likely to be measureable.

A SYSTEM-WIDE EUTROPHICATION MODEL OF NEW YORK HARBOR, LONG ISLAND SOUND AND NEW YORK BIGHT

John P. St. John, P.E.
Charles L. Dujardin, P.E.
Robin L. Miller
HydroQual, Inc.
1 Lethbridge Plaza
Mahwah, New Jersey 07430

Warren Kurtz, P.E.
Robert Gaffoglio, P.E.
New York City Department of Environmental Protection
Bureau of Environmental Engineering

ABSTRACT

The City of New York is conducting facility upgrade planning for the Newtown Creek Water Pollution Control Plant (WPCP), located on the East River in Brooklyn, New York. The objective of the planning project is to develop viable management options for treated sewage effluent discharged from this facility and from other plants located on the East River. As part of this planning, the City authorized the development and application of the System-Wide Eutrophication Model (SWEM). The purpose of SWEM is to provide refined evaluations of the effects of proposed WPCP upgrades and other engineering alternatives and, ultimately, to select a cost-effective management plan for the East River and western Long Island Sound.

New York City is proceeding to achieve secondary treatment at the Newtown Creek WPCP. In addition, the Long Island Sound Study of the National Estuary Program is requiring that New York City abate the contribution of nitrogen loadings from the six East River WPCPs to alleviate low dissolved oxygen levels in western Long Island Sound. Engineering solutions under consideration to satisfy these mandates include: advanced treatment at the East River WPCPs, aggregation of East River discharges, outfall relocation, and tidal gates. The nature of the engineering solutions necessitated that SWEM include both detailed hydrodynamics and eutrophication kinetics. Nearly 30 engineering alternatives were evaluated. This paper summarizes the water quality issues, technical approach, and results for three diverse engineering solutions.

KEYWORDS

New York Harbor, Long Island Sound, New York Bight, Eutrophication, Dissolved Oxygen, Nitrogen Reduction, Outfall, Tidal Gates

Abstract Southeastern Connecticut and Distant Places

A Comparison of the Effects of Acid Rain

by Erwin M. Cohen

From a continuing study of the acidification of water supplies and soils in Southeastern Connecticut, the effects of acid rain on other areas of the Eastern Seaboard and on central and northern Europe have been compared. Although acid rain has continued to fall on the entire East coast of the United States for 30 years. the effects are often more noticeable where specific environmental conditions exist. These conditions include the buffering capacity of local soils. From the alkaline soils of the Delaware Basin to the granitic soils of Southeastern Connecticut, the White Mountains of New Hampshire, and the Adirondacks of New York the soils of the region play a significant part in what effects the acid rain will have. So also in Sweden and Norway where the soils resemble those in Southeastern Connecticut.

As a result of local conditions the effects of acid rain can be very specific. These include acidification of water supplies and subsequent contamination by trace metals, damage to selected forests areas resulting in stunted growth, and leaching of toxic metals from both natural and industrialized area into streams ponds and rivers.

Only a few countries such as Sweden, have taken positive steps to protect the health and property of their citizens by directly providing grants for treatment of acid waters. In the United States reliance on acid trading credits has not protected sensitive regions because often trades are made that allow the acid rain to continue to fall on the sensitive regions.

The paper will present tables and charts of local conditions and acid rain effects from personal investigations and from the published literature.

Affiliation EMCO Testing and Engineering Company

P.O. Box 644

Norwich, CT 06360 E-Mail EMCO@99main.com

CHAOTIC ADVECTION AND ITS ROLE IN THE WATER QUALITY OF THE LOWER HUDSON RIVER ESTUARY

Amvrossios C. Bagtzoglou, Andrei Novikov, Brandon Gallagher, and Rene Chevray

School of Engineering and Applied Science
Columbia University
New York, NY 10027, USA

Introduction

In the present study we are dealing with the control of the Lower Hudson River Estuary (LHRE) water quality. As estuary can be defined the confluence of a freshwater river with salt water, where mixing occurs under tidal action. Tidal waters are receiving much recognition nowadays since greater quantities of pollution are moving towards the mouth of larger rivers. Many of these large rivers or streams terminate at the ocean where they are subjected to the effect of tidal currents. Estuaries are very sensitive and their ecosystems can be very easily upset. The US Clean Water Restoration Act of 1966 defines "estuarine zones", as environmental systems consisting of an estuary and those transitional areas that are consistently influenced or affected by water from an estuary, such as, but not limited to, salt marshes, coastal and intertidal areas, bays, harbors, lagoons, in-shore waters and channels. The LHRE is a partially enclosed area where the tidal actions of the Atlantic Ocean mix water from rivers, streams and runoff with salt water. As this system neighbors the highly populated urban areas of New York, New Jersey and Connecticut, as well as zones with strong industrial activity it is subject to pollution. A specific example is the release of polychlorinated biphenyls (PCBs) in 1970's from two General Electric (GE) capacitor-manufacturing plants located in Fort Edward and Hudson Falls, NY. Even though many of GE's PCBs remain concentrated in hotspots in the sediments of the upper Hudson, PCBs have polluted the entire stretch of Hudson River below Hudson Falls. Environmental Protection Agency (EPA) studies in 1997 not only showed that the problem exists 30 years later but they also proved that the PCB load would not be naturally remediated via dechlorination. Such a fact underlines, the great importance of the search for and development of alternative remedial options.

Chaotic Advection

It has long been recognized that particle paths in a smooth laminar flow may be chaotic. This phenomenon where a flow is laminar and some particle paths are chaotic is called *Lagrangian turbulence* but we rather prefer the name *chaotic advection* established by Aref (1984) because the flow is in fact not turbulent. When discrete particles are introduced into a chaotically advected medium, they are often distributed rapidly over the accessible space. Under certain conditions it is also possible that the particles entrained in this flow become segregated to a specific region, causing that way local inhomogeneity or changes in the boundary geometry. This kind of processes could be observed in many natural phenomena such as the action of tidal waves and tidal currents in particle and sediment transport. There, the incoming and outgoing tides generate periodic streamline patterns, which could induce chaotic Lagrangian trajectories.

Particle transport in such conditions could be simulated with the process of industrial mixing or of dispersion of pollutants in natural media.

In general, mixing consists of stretching and folding of fluid filaments, and distribution throughout space accompanied by breakup if the fluids are sufficiently different. Alternatively, mixing may be said to be a combination of *stirring*, *mass diffusion across intermaterial surfaces* and *possibly chemical reactions*. Stirring is the stretching and distribution of fluid elements throughout space, which leads to the creation of an intermaterial area and it reduces the length scales between the initially segregated fluids. By producing more intermaterial area stirring promotes mixing (Aref and Balachandrar 1986). Thus, mixing depends on material parameters, for example, diffusivities, whereas stirring depends only on parameters governing the flow. As a result one can stir fluids that do not mix at all. The simplest mathematical model of stirring appears when particles are advected passively by a prescribed or evolving velocity field.

Chaotic Advection in Open Water Bodies

Even though there exist numerous studies documenting chaotic advection phenomena in closed flows (Chaiken et al., 1986; Chien et al., 1986; Jones, 1991; Dutta and Chevray, 1995), work in this field has only recently started involving open water bodies. An important case of natural flows is the flow representative of tidal areas, which are characterized by their shallowness and their strong tidal currents. The nonlinear interaction of tidal currents and bottom topography produces currents, which give nonzero contributions to the tidally averaged currents. These residual currents are important for the transport and mixing properties of the tidal flow. Initially the flow in tidal areas can be considered to be the superposition of a time dependent part (tidal flow) and a time independent part (residual flow). Pasmanter (1988) first mentioned that the advection problem in this type of flows could be expressed as perturbed two-dimensional (2D) Hamiltonian systems. This type of system could be chaotic. Pentek et al. (1999) have also investigated passive advection in open flows where the time-dependent region of the flow is assumed to be restricted to a finite domain called the *mixing region*.

Ridderinkof and Zimmerman (1992) studied particle motion in tidal areas using a kinematic model, which can be interpreted as a strongly perturbed Hamiltonian system. Beerens et al. (1994) considered a dynamical model for tide topography interaction. Lagrangian advection in this flow can be described in terms of perturbed Hamiltonian systems. Their numerical results showed that the method works only for very small perturbations. In case of strong perturbations (tidal areas with irregular bottom topography) the method of orbit expansion could be applied giving an approximation of a Poincare section. Using this approximation, displacements near the cell boundary can be studied and chaotic mixing coefficients can be calculated.

Ridderinkof and Zimmerman (1992) have analyzed a 2D tidal model of the Wadden Sea (a tidal area located behind a chain of islands off the Dutch, German and Danish coasts) revealing Lagrangian chaos in the trajectories of water. Although the Eulerian velocity field in the Wadden Sea was much more complicated than most laboratory and numerical models studied before, Ridderinkof and Zimmerman (1992) were able to prove that particle spreading was similar to chaotic stirring in simple deterministic time periodic 2D velocity fields. Thus, the large spreading of particles in the Wadden Sea can be attributed to the presence of chaotic stirring. So the presence of chaotic trajectories in the Wadden Sea and Western Scheldt estuary has been confirmed. Kranenburg (1992) has also shown that wind-driven chaotic advection is possible in a shallow model lake. Finally, several recently reported research efforts (Yang and Liu, 1997; Toroczkai et al., 1998; Tamas et al., 2000) make it clear that the field of chaotic advection is nowadays concentrating on open water body applications.

Present Study

Our aim with the present work is to prove the existence of chaotic advection in the LHRE system with the implementation of a 2D (depth averaged) hydrodynamic numerical model. The model is based on the diffusion approximation of the Saint-Venant equations and has been calibrated and tested with the help of NOAA data. Our flow results are fed into particle tracking transport simulations that will allow us to test the hypothesis that this estuarine system is behaving like a chaotically advected system. If this hypothesis were to be proven true, an explanation for the relatively good water quality found in the LHRE system may be attained.

References

1. Aref H., and S. Balachandar, Chaotic advection in a Stokes flow, Phys. Fluids, vol. 29(11), pp.3515-3521, 1986
2. Aref H., Stirring by Chaotic advection, J. Fluid Mech., Vol. 143, pp.1-21, 1984
3. Beerens S.P., H. Ridderinkhof and J.T.F. Zimmerman, An analytical study of chaotic Stirring in Tidal Areas, Chaos, Solitons and Fractals, vol. 4(6), pp. 1011-1029, 1994
4. Chaiken J., R. Chevray, M. Tabor, and Q.M. Tan, Experimental study of Lagrangian turbulence in a Stokes flow, Proc. R. Soc. Lond. A, vol. 408, pp. 165-174, 1986
5. Chien W.L., H. Risisng, J.M. Ottino, Laminar mixing and chaotic mixing in several cavity flows, J.Fluid Mech. Vol. 170, pp. 355-377, 1986
6. Dutta P., and Rene Chevray, Inertial effects in chaotic mixing with diffusion, J. Fluid Mech. Vol. 285, pp. 1-16, 1995
7. Jones S., The enhancement of mixing by chaotic advection, Phys. Fluids A 3(5), pp.1081-1086, 1991
8. Kranenburg C., Wind-driven chaotic advection in a shallow model lake, Journal of Hydraulic Research, vol. 30(1), pp. 29-46, 1992
9. Pasmanter R.A., Deterministic diffusion, effective shear and patchiness in shallow tidal flows, Physical Processes in Estuaries, pp. 42-52, Springer, 1988
10. Pentek A., G. Karolyi, I. Scheuring, T. Tel, Z. Toroczkai, J. Kadtke, C. Grebogi, Fractality, chaos and reactions in imperfectly mixed open hydrodynamical flows, Physica A, vol. 274, pp.120-131, 1999
11. Ridderinkhof H., and J.T.F. Zimmerman, Chaotic stirring in a tidal system, Science, Vol. 258, pp.1107-1111, 1992
12. Tamas T., G. Karolyi, A. Pentek, Z. Toroczkai, C. Grebogi, J. Kadtke, Chaotic advection, diffusion, and reactions in open flows, Chaos, Vol. 10 (1), pp. 89-98, 2000
13. Toroczkai Z., G. Karolyi, A. Pentek, T. Tel, C. Grebogi, Advection of active particles in open chaotic flows, Physical Review Letters, vol. 80(3), pp. 500-503, 1998
14. Yang H., Z. Liu, The three-dimensional Chaotic Transport and the Great Ocean barrier, Journal of Physical Oceanography, vol. 27 pp. 1258-1273, 1997

THE ROLE OF KEROGEN IN THE SORPTION OF HYDROPHOBIC ORGANIC CHEMICALS BY A SANDY AQUIFER MATERIAL

WEILIN HUANG
School of Environmental Science, Engineering and Policy
Drexel University, Philadelphia, PA 19104

YONG RAN
Institute of Geochemistry, Chinese Academy of Sciences
Wushan, Guangzhou, People's Republic of China

BAOHUA XIAO
School of Environmental Science, Engineering and Policy
Drexel University, Philadelphia, PA 19104

JIAMO FU
Institute of Geochemistry, Chinese Academy of Sciences
Wushan, Guangzhou, People's Republic of China

GUOYING SHENG
Institute of Geochemistry, Chinese Academy of Sciences
Wushan, Guangzhou, People's Republic of China

INTRODUCTION

Sorption by soils and sediments is a phase distribution process affecting chemical and biological reactivity, transport, and risks of toxic chemicals in aquatic environments. It is known that natural organic matter (NOM) associated with soils and sediments is the dominant constituent for sorption, sequestration and attenuation of relatively hydrophobic organic chemicals (HOCs) in aquatic systems [1-5]. The chemical heterogeneity of NOM also affects the rates and equilibria of sorption and desorption of HOCs in soils and sediments [5-7]. A concept of "soft carbon" (or amorphous, rubbery state) vs. "hard carbon" (or condensed, glassy state) NOM has been invoked to operationally delineate chemical heterogeneity of NOM and to elucidate the mechanisms for sorption by soils and sediments [5-10]. In the context of this conceptual model, the amorphous NOM or aNOM domains follow a partitioning process and the condensed NOM or cNOM domains follow external/internal surface adsorption processes [5-10]. It is hypothesized that cNOM dominates isotherm nonlinearity, sorption-desorption hysteresis, slow rates of sorption and desorption, and competition among coexisting HOCs exhibited by soils and sediments [5-10]. Identification of cNOM and quantification of its contents in soils and sediments are thus critical to appropriate interpretation and accurate prediction of varied sorption and desorption phenomena.

This study demonstrates that, even present at trace levels, cNOM dominates the overall sorption behavior exhibited by soils and sediments having low total organic carbon (TOC) contents. It is widely held that mineral surfaces may play a significant role in the uptake and sequestration of moderately HOCs if a soil or sediment contains TOC < 0.1 wt% [2, 3, 10-13]. In this study we challenged this hypothesis by examining the dominant role of the cNOM in the sorption of several HOCs for a low TOC sandy aquifer material collected from Borden, Ontario, Canada. Borden material was well characterized and was utilized as a sorbent in several HOC sorption rate and equilibrium studies [11-15]. The results of these prior studies showed that the rates of HOC uptake by the material were very slow, and that the measured sorption distribution coefficients (K_D), the ratios of the sorbent-phase concentration (q_e) to the aqueous phase concentration (C_e) of several HOC solutes, were greater than their respective K_D values predicted from hydrophobicity (i.e., K_{OW}) of the HOCs and the organic carbon fraction (f_{OC}) based on empirical log K_{OC} – log K_{OW} correlations. The observed overall slow rates of sorption were attributed to retarded intraparticle diffusion and the higher-than-predicted sorption coefficients were attributed to binding of HOC solutes to mineral surfaces [11-14].

In this study we isolated and identified kerogen-like cNOM from the Borden material. Sorption isotherms were measured for both the original sand and the isolated cNOM using four different HOC as the target solutes. Our study showed that, even though its content is very low, the condensed kerogen material dominates the overall isotherm nonlinearity, sorption capacities and sorption-desorption hysteresis of the subsurface material.

EXPERIMENTAL SECTION

Sorbents and Sorbent Characterization

Borden aquifer material consists mainly of mineral grains and rock fragments of sizes ranging from about ~850 to ~75 μm. Carbonate rock fragments are about 13~17 wt% of the bulk material, which contain minerals of biological origins (i.e., biomicrite and biosparite) and the majority of NOM of the aquifer material. The sand has low TOC (~ 0.021 wt%), low specific surface area (~0.4 m^2/g), and low pore volume (<0.009 cm^3/g) [12-15].

NOM was isolated from the sand using an acid (HCl) digestion procedure. After decarbonation, fine particles of the treated material were separated for further demineralization using concentrated HCl + HF acids. The residue left after demineralization was sequentially washed with HCl acid and Milli Q water and then was oven dried at 65°C. The dried residue was split to two portions; one was stored in a glass bottle for use in sorption experiments and the other was further treated with hot concentrated hydrochloric acid to remove fluoride and was used for NOM characterization.

The isolated NOM were analyzed for its elemental composition using a high temperature combustion method on a Heraeus CHN-O elemental analyzer.

Functionality of the second fraction of the isolate was determined using solid-state CP/MAS ^{13}C-NMR on a 7.5-mm probe at 50 MHz carbon frequency. The organic facies and the shapes, sizes, degree of maturation of the second portion of the isolated SOM were examined under optical microscopy. Before optical characterization, the NOM sample was mixed with a liquefied organic resin and the mixture was mounted on glass slides. After solidification, the mixture mounted were sectioned, polished and examined under optical microscopy in both transmitted and reflected modes and fluorescence microscopy (Leitz MPV3) with blue light as light source. Fluorescence measurements were made under conditions of 100 W ultra-high-pressure Hg-lamp, "L" 2 blue light excitation (wavelength 435 nm) using PLOEMOPAK 2, beam splitter 510 nm and barrier 490 nm. Vitrinite reflectance was measured using interference green filter (546 nm), diaphragm (5x), and sapphire as reflectance standard.

Equilibrium Sorption and Desorption Experiments

Phenanthrene (Phen), naphthalene (Naph), 1,3,5-trichlorobezene (TCB), and 1,2-dichlorobezene (DCB), obtained in spectrophotometric grade (>98%) from Aldrich Chemical Co., Inc., were used as the HOC probe solutes for this study. The octanol-water partition coefficients (log K_{ow}), aqueous solubilities (C_S), and the K_{OC} values estimated from empirical correlations reported in two different references [3,4] are listed in Table 1. Sorption and desorption equilibria were measured for both the original sand and the first portion of the isolated NOM using flame-sealed glass ampules as batch reactors. Laboratory procedures published elsewhere [5-7, 16, 17] were followed exactly.

TABLE 1. MAJOR PROPERTIES OF THE SELECTED ORGANIC CHEMICALS

Chemical	Molecular Wt (g/mol)	log K_{OW}	C_S (mg/L)	K_{OC} (ml/g)	
				[3]	[4]
Phenanthrene	178.2	4.57	1.12	22900	17400
Naphthalene	128.2	3.3	31.7	1230	1750
1,2,3-TCB	181.5	4.02	5.36	6460	5930
1,2-DCB	147.2	3.38	92.9	1480	840

RESULTS AND DISCUSSION

Organic Matter Characterization

The isolated material is a gray colored powder. The first portion used in sorption experiments contains only ~ 0.736 wt% of TOC and the second portion contains 5.56 wt% of TOC. X-ray diffraction and petrographical examinations showed that fluorite formed during HF/HCl treatment is the dominant component of the isolated NOM.

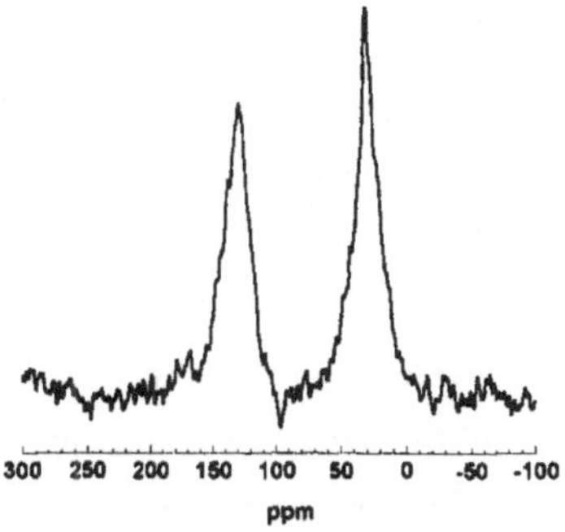

Figure 1. CP/MAS ^{13}C-NMR spectrum for the isolated/purified NOM (TOC = 5.7 wt%).

^{13}C NMR spectrum shown in Figure 1 has two major peaks at ~33 and ~137 ppm representing aliphatic (0~60 ppm, peak area ~53.4%) and aromatic carbons (100~160 ppm, peak area ~ 46.6%), respectively. This is typical for diagenetically altered and chemically reduced kerogen materials [5].

Examinations under fluorescence and optical microscopes show that the isolated NOM is highly heterogeneous at the particle scale. The organic particles are irregularly shaped and have sizes ranging from sub-micron to about 40 μm. Both vitrinite and bituminite or liptinite are the dominant macerals, and fusinite is a minor component found in the isolated NOM. Vitrinite is likely originated from plant materials and is the major component of Kerogen Type III. Liptinite or bituminite is commonly derived from animal's tissues or lower level plants such as algae and is the major component of Kerogen Type I and II. Fusinite, also called black carbon or charcoal, is typically formed during incomplete oxidation of Kerogen Types I~III and is classified as Kerogen Type IV. Vitrinite found in the isolate is opaque under transmitted microscopy, not fluorescent and is dark-grey and dark colored under fluorescence microscopy. Liptinites are found semi-transparent under optical microscopy and are yellowish green to yellow and brownish colored and manifest amorphous character under fluorescence microscopy. The reflectance (Ro) measured for vitrinite and bituminite are about 0.66 ~ 0.82 % and 0.1 ~ 0.23%, respectively, indicating relatively low maturation of the isolated kerogen material. Fusinite and semifusinite are opaque (transmitted), have much higher reflectance (reflected), are not fluorescent (fluorescence).

Figure 2. Tubular vitrinite (b), amorphous bituminite (d) and fusinite (g) under fluorescence, transmitted and reflected microscopes, respectively.

Sorption and Desorption Isotherms and Hysteresis

The sorption and desorption equilibrium data obtained for both sandy material and the isolated NOM (TOC = 0.735wt%) were fitted to the Freundlich isotherm model having a form of

$$q_e = K_F C_e^n \quad (1)$$

or

$$\log q_e = \log K_F + n \log C_e \quad (2)$$

Here q_e and C_e are the solid-phase and aqueous-phase equilibrium solute concentrations, expressed in units of μg/g and μg/L, respectively; the parameter K_F and the exponent n are the Freundlich isotherm capacity parameter and isotherm linearity factor, respectively. Linear regression procedure using SYSTAT software (Version 8.0, SYSTAT Inc.) was utilized for fitting Eq 2 to sorption and desorption data. Representative sorption and desorption data for two of the eight sorbent-sorbate systems studied are shown in Figures 3 and 4.

Our results show that sorption and desorption isotherms are nonlinear, that sorption coefficients are greater than those predicted from the reported log K_{OC} - log K_{OW} correlations, and that sorption and desorption are hysteretic. The modeling exercises indicate that the Freundlich equation fits the isotherm data well and that the values of the fitted Freundlich parameter n range from 0.69 to 0.88 for all but one sorbent-sorbate system (Figure 4). A plot of isotherm linearity (n) versus hydrophobicity of the HOC solutes (log K_{OW}) yields an inverse linear relationship, indicating that the more hydrophobic the HOC solute is, the more nonlinear its sorption isotherm (Figure 5).

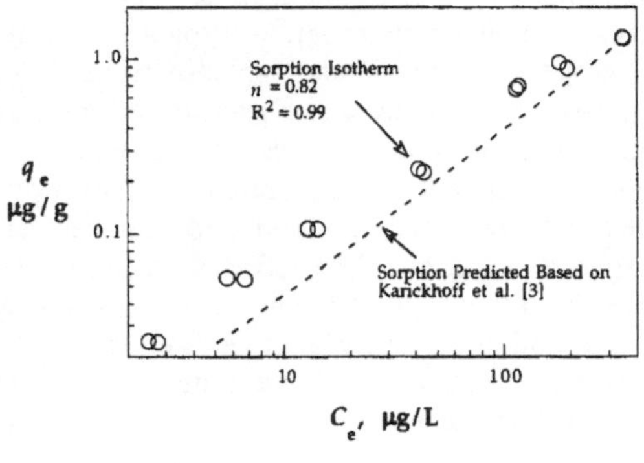

Figure 3. Phenanthrene sorption isotherm measured for the original sand. n and R^2 are the isotherm linearity and the goodness of fitting, respectively. The dashed line represents phenanthrene sorption equilibrium predicted from the log K_{OC} - log K_{OW} correlation reported in [3].

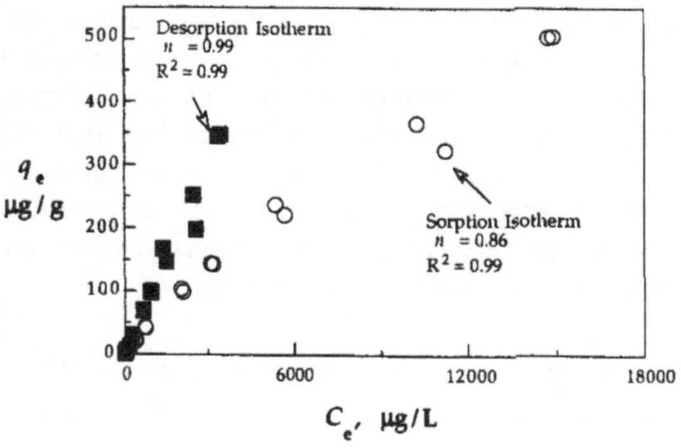

Figure 4. Naphthalene sorption and desorption isotherms measured for the isolated NOM. n and R^2 are the isotherm linearity and the goodness of fitting, respectively.

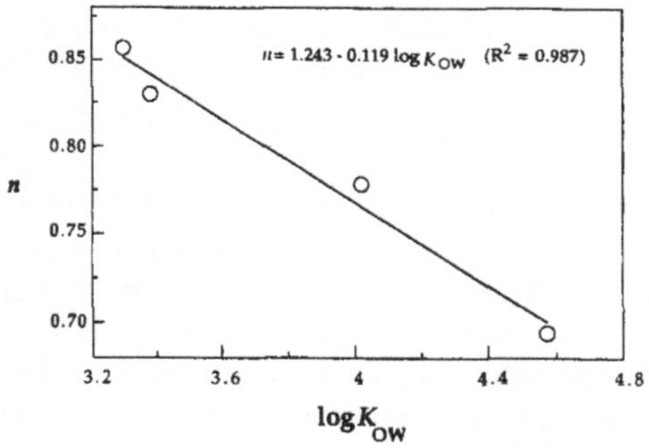

Figure 5. Isotherm linearity (n) measured for the isolated NOM as a function of log K_{OW} value of the HOC solutes.

The nonlinear, hysteretic and high-capacity sorption-desorption phenomena observed for Borden material apparently result from the presence of relatively reduced and condensed kerogen material. As shown in Figure 1 the isolated NOM material has only two oxygen-deficient functional groups: aliphatic and aromatic hydrocarbon groups. This highly hydrophobic nature of the kerogen particles greatly increases the organic carbon normalized sorption capacities.

Our data indicates that the measured single point K_{OC} values for the original aquifer material are greater than their respective values predicted from log K_{OC} – log K_{OW} correlations (Figure 6). This is consistent with the observations reported previously for the same material. While previous studies attributed this to the HOC sorption by mineral surfaces, our results clearly indicate that it is due to the presence of condensed NOM in the sand. It is apparent from Figure 6 that the organic carbon normalized single-point sorption coefficient, K_{OC}, for the isolated cNOM changes as a function of both aqueous-phase solute concentration (C_e) and the hydrophobicity of the solute (K_{OW}); the C_e dependence of K_{OC} is a result of isotherm nonlinearity. The K_{OC} values reported in this study for the isolated NOM are several times greater than their respective values predicted from the widely used log K_{OC} – log K_{OW} correlations such as those reported in ref [2-4]. Linear regressions for the K_{OC} values calculated at three different C_e levels using the fitted isotherm parameters yield the following three empirical equations for the isolated kerogen material.

$$\log K_{OC} = 1.348 \log K_{OW} - 0.459 \qquad C_e @ 0.001\, S_w \quad R^2 = 0.991 \qquad (3)$$
$$\log K_{OC} = 1.151 \log K_{OW} - 0.065 \qquad C_e @ 0.05\, S_w \quad R^2 = 0.991 \qquad (4)$$
$$\log K_{OC} = 1.008 \log K_{OW} + 0.239 \qquad C_e @ 0.75\, S_w \quad R^2 = 0.988 \qquad (5)$$

A simple calculation can also show that sorption on mineral surfaces is insignificant for this sandy material when compared to the uptake by cNOM. In ref [18] we reported a mass-based phenanthrene K_D value of 1.2×10^{-3} L/g for a natural quartz sample (N_2-BET SSA = 6.53 m^2/g). Given the SSA of 0.4 m^2/g for Borden material, the contribution from mineral surfaces to the phenanthrene K_D value could be about 7.35×10^{-5} L/g, which is about 1~2 percent of the overall phenanthrene K_D value ($\geq 4\times10^{-3}$ L/g) measured for the bulk sand.

It appears that sorption by this low TOC material, one of the most homogeneous aquifer materials, is very complicated. The three different types of kerogen particles found in the sand (Figure 2) likely exhibit very different sorption phenomena. It is possible that sorption into the amorphous bituminite material follows a partitioning process whereas sorption to the more condensed and matured vitrinite and fusinite particles follows surface adsorption process. The carbonate mineral matrices that encapsulate these kerogen particles may reduce the accessibility of the NOM to HOC solute molecules during sorption process. This could reduce the rates of sorption and increase apparent sorption-desorption hysteresis for the bulk sandy material.

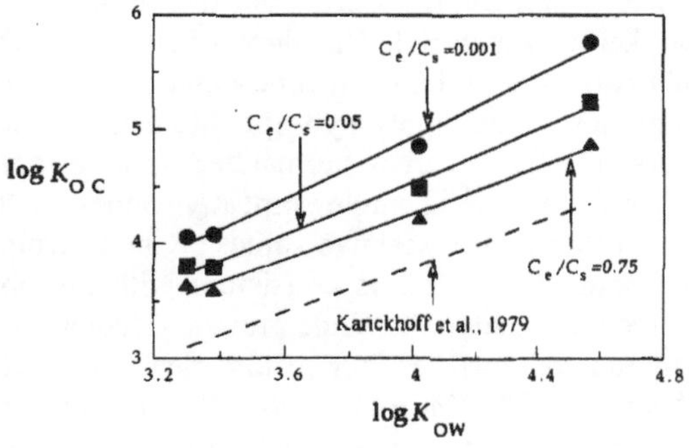

Figure 6. log K_{OC} value (in units of ml/g) of the isolated NOM as functions of aqueous-phase concentration and log K_{OW} value of HOC solutes.

REFERENCES

1. Chiou, C. T., L. J. Peters, and V. H. Freed. *1979*. "A Physical Concept of Soil-Water Equilibria for Non-ionic Organic Compounds," *Science*, 206: 831-832.

2. Karickhoff, S. W. *1984*. "Organic Pollutant Sorption in Aquatic Systems," *J. Hydraulic Engin.*, 110: 707-735.

3. Karickhoff, S. W., D. S. Brown, and T. A. Scodd. *1979*. "Sorption of Hydrophobic Pollutants in Natural Sediments," *Wat. Res.*, 13: 241-248.

4. Means, J. C., S. G. Wood, J. J. Hassett, and W. L. Banwart. *1980*. "Sorption of Polynuclear Aromatic Hydrocarbons by Sediments and Soils," *Environ. Sci. & Technol.*, 14(12): 1524-1528.

5. Huang, W., and W. J. Weber, Jr. *1997*. "A Distributed Reactivity Model for Sorption by Soils and Sediments. 10: Relationships Between Sorption, Hysteresis, and the Chemical Characteristics of Organic Domains," *Environ. Sci. & Technol.*, 31(9): 2562-2569.

6. Huang, W., and W. J. Weber, Jr. *1998*. "A Distributed Reactivity Model for Sorption by Soils and Sediments. 11: Slow Concentration-dependent Sorption Rates," *Environ. Sci. & Technol.*, 32(22): 3549-3555.

7. Johnson M. D., W. Huang, Z. Dang, and W. J. Weber, Jr. *1999*. "A Distributed Reactivity Model for Sorption by Soils and Sediments. 12: Partial Removal of Amorphous Organic Domain and Its Effect on Sorption Equilibria," *Environ. Sci. & Technol.*, 33(10): 1657-1663.

8. Weber, W. J., Jr., P. M. McGinley, and L. E. Katz. *1992*. "A Distributed Reactivity Model for Sorption by Soils and Sediments: 1. Conceptual Basis and Equilibrium Assessments," *Environ. Sci. & Technol.*, 26: 1955-1962.

9. Xing, B., and J. J. Pignatello. *1997*. "Dual-Mode Sorption of Low-Polarity Compounds in Glassy Poly(Vinyl Chloride) and Soil Organic Matter," *Environ. Sci. & Technol.*, 31(3): 792-799.

10. LeBoeuf E. J., and W. J. Weber, Jr. *1997*. "A Distributed Reactivity Model for Sorption by Soils and Sediments. 8. Identification of A Humic Acid Glass Transition and A Logic for Invoking Polymer Sorption Theory." *Environ. Sci. & Technol.*, 31:1697-1702.

11. Curtis, G. P., M. Reinhard, and P. V. Roberts. *1986*. "Sorption of Hydrophobic Organic Compounds by Sediments," In *Geochemical Processes at Mineral Surfaces* (Edited by Davis, J. A., and K. F. Hayes), American Chemical Society, Washington DC.

12. Ball, W. P., and P. V. Roberts. *1991*. "Long-Term Sorption of Halogenated Organic Chemicals by Aquifer Materials: Part 1. Equilibrium," *Environ. Sci. & Technol.*, 25: 1223-1237.

13. Ball, W. P., and P. V. Roberts. *1991*. "Long-Term Sorption of Halogenated Organic Chemicals by Aquifer Materials: Part 2. Intraparticle Diffusion," *Environ. Sci. & Technol.*, 25: 1237-1249.

14. Curtis, G. P., P. V. Roberts, and M. Reinhard. *1986*. "A Natural Gradient Experiment on Solute Transport in a Sand Aquifer: 4. Sorption of Organic Solutes and Its Influence on Mobility," *Wat. Resour. Res.*, 22: 2059-2067.

15. Ball, W. P., C. Buehler, T. C. Harmon, D. M. Mackay, and P. V. Roberts. *1990*. "Characterization of a Sandy Aquifer Material at the Grain Scale," *J. Cont. Hydrol.*, 5: 253-295.

16. Weber, W. J., Jr., and W. Huang. *1996*. "A Distributed Reactivity Model for Sorption by Soils and Sediments: 4. Intraparticle Heterogeneity and Phase-Distribution Relationships under Non-equilibrium Conditions," *Environ. Sci. & Technol.*, 30(3): 881-888.

17. Huang, W., Yu, H., and W. J. Weber, Jr. *1997*. "Hysteresis in the Sorption and Desorption of Hydrophobic Organic Contaminants by Soils and Sediments: 1. A Comparative Analysis of Experimental Protocols," *J. Cont. Hydrol.*, 31: 129-148.

18. Huang, W., M. A. Schlautman, and W. J. Weber, Jr. *1996*. "A Distributed Reactivity Model for Sorption by Soils and Sediments: 5. The Influence of Near-Surface Characteristics in Mineral Domains," *Environ. Sci. & Technol.*, 30(9): 2993-3000.

SYMPOSIUM 6:

Environmental Management and GIS

DEVELOPING AND IMPLEMENTING ENVIRONMENTAL MANAGEMENT SYSTEMS

ADAM H. STEINMAN, ESQ.
Woodard & Curran
41 Hutchins Drive
Portland, ME 04102

MARY HOUSE
Woodard & Curran
41 Hutchins Drive
Portland, ME 04102

MARYKRISTIN IVANOVICH
Woodard & Curran
980 Washington Street
Suite 325N
Dedham, MA 02026

ABSTRACT

As EPA enforcement against and the number of rules applicable to colleges and universities continue to rise, the ability of an EHS manager to maintain a college campus in compliance with rules becomes increasingly more difficult. As a result of these pressures, more and more campuses are starting to develop and implement Environmental Management Systems (EMS). Several international standards for EMS development have served as the basis for EMS including the ANSI/ISO 1400, which was developed in 1996 to address the need for a standard against which EMS could be measured. This standard offers a complete and internationally recognized process for EMS development and is inherently flexible in allowing different types and sizes of organizations to focus on issues they identify as most important. Organizations can develop an EMS that is ISO 14001 certifiable, or they can tailor a system to address their specific needs using a practical approach that borrows from the international standard.

An effective EMS provides an organized systematic approach for the development, implementation, review and continual improvement of a campus', company's or facility's environmental management program. Before embarking on the EMS, a facility must clearly define the purpose, scope and objectives of the EMS. An EMS consists of three distinct phases: planning; implementation; and continual improvement. Each phase requires the completion of several substantial activities. For example, identifying environmental aspects and detailing a facility's obligations to the applicable environmental regulations is a major activity in the planning phase of an EMS.

This paper provides the background of the ISO 14001 standard, explains the benefits of developing an EMS, gives an overview of the various developmental phases, and describes tools that companies can use to facilitate the planning, implementation and monitoring of the EMS. It underlines the importance of resource allocation, management support and personnel ownership for the successful execution of an organization's EMS.

Dusting for Plume Footprints: Using Atmospheric Dispersion Modeling and GIS To Assess The Environmental Impact of Particulate Emissions From the Con Edison Ravenswood Generating Station,

Jeffrey P. Osleeb, Ph.D., Program in Earth and Environmental Sciences Graduate Center - City University of New York, New York, NY, josleeb@gc.cuny.edu and Joy Romanski, Department of Earth and Environmental Sciences, Columbia University, New York, NY

Abstract

Current research conducted by faculty and students at the University of Ruhr and at Hunter College indicates the presence of an unusually high amount of industrial pollutants (particularly fly ash) at the N and the R subway platforms at the 59th Street and Lexington Avenue station. Given the relatively industrially pristine nature of the Upper Eastside of Manhattan, the authors determined that the source of these pollutants must be at some distance from the station and are somehow being transported to the station.

A geographic information systems (GIS) investigation was conducted to determine all potential sources of this pollution. After surveying a wide area in which some three hundred potential sources were screened, the suspected source of the pollutants appears to be the Con Edison (now owned by Keyspan) Ravenswood Generating Station in Long Island City, New York. However, the Ravenswood Generating Station is over one mile away from the station.

However, it is also recognized that pollution is typically transported over significant distances. This transport of pollutants is impacted by local weather conditions, the built environment and the nature of the pollutants. Therefore, this study employs geographic plume analysis (GPA) techniques to model the dispersion of particulate emissions from the generating station and thereby determines the potential sites of deposition of these particulates.

Potential sites for the entry of these pollutants into the subway station are investigated. In addition, a mechanism for the movement of the pollutants within the tunnel is hypothesized. In this way the impact on air quality within the subway tunnel system by the Ravenswood Generating Station is determined.

Integration with GIS of Large-Scale Water Quality Projects

In the New York City Area

Gary Ostroff, P.E. Sr. Project Manager
HydroQual, Inc. 1 Lethbridge Plaza, NJ 07430
www.hydroqual.com

In the course of our project work for the City of New York and other organizations, HydroQual has created many mathematical models of the region's major water bodies. In addtion, we have pursued many city-wide planning goals to address water pollution issues. The models generally link together our proprietary hydrodynamic and water quality models to produce a complete simulation of water movement and reaction kinetics within the subject area. With the advent of desktop GIS, we are moving to fully integrate our modeling work into a GIS environment in order to take advantage of the tremendously rich spatial databases available for the NYC area. We have also realized tremendous gains in efficiency in setting up models by employing GIS as a pre-processing tool.

The current effort to create a comprehensive NYC base map promises to bring about a revolution in the use of GIS in the New York area, but the existing map, COGIS, affords a highly detailed, if not geographically precise, view of properties throughout the five boroughs. For our work in the area of water quality, this spatial data has been invaluable for locating sites, estimating catchment loading, assessing catchment characteristics of importance for water quality, and creating myriad presentation maps, on paper and on the Web.

Before this integration effort began, our modeling work was completely separate from our GIS work, and model output was displayed in black and white graphs or animations. In either case, it was not possible to directly relate the model results to digital map data that is available to users of GIS. HydroQual is now pursuing ways to animate model output within the GIS environment using specialize programs or available extensions. The investigations required for this work returned us to the consideration of may interesting and fundamental questions about the representation of multi-variable data in maps, digital or traditional.

SYMPOSIUM 7:

Environmental Molecular Engineering

INTEGRATED COMPUTATIONAL AND EXPERIMENTAL INVESTIGATION OF THE THERMODYNAMIC PROPERTIES OF POLYCHLORINATED BIPHENYLS

SWATI PURI, WILLIAM J. WELSH AND JAMES S. CHICKOS
Department of Chemistry and Biochemistry
University of Missouri-St. Louis
8001 Natural Bridge Road
St. Louis, Missouri 63121

INTRODUCTION

Among the environmental chemicals that may be able to disrupt the endocrine system of animals and humans, polychlorinated biphenyls (PCBs) have attracted considerable attention. PCBs, a class of chlorinated hydrocarbons, may interfere with endocrine function due to their ability to mimic natural hormones. The lipophilic nature of these organohalogen compounds has led to their accumulation in the food chain, and it is the cause of adverse human health effects, such as, chloracne and related dermal lesions, diverse hepatic effects, and decreased pulmonary function. PCBs entered the environment as components of pesticides, plasticizers and adhesives. The same physicochemical properties of PCBs, namely non-flammability and chemical stability, which contributed to the widespread use of these compounds in industrial applications, are also partially responsible for their designation as organic pollutants.

PCBs have been found at an appreciable concentration in the polar region, presumably as a result of long-range atmospheric transport. For many environmental applications, the subcooled liquid vapor pressures and vaporization enthalpies are the relevant thermodynamic properties associated with the dispersal of PCBs. For a particular PCB isomer, its mode and rate of migration through the global atmosphere depends in large part on its volatility. Highly volatile PCBs tend to remain airborne and migrate faster. Less volatile PCBs are more prone to partition into soil or vegetation.

A study of the thermodynamic properties of PCBs is essential to assess the impact of these compounds on the environment. The magnitude of the fusion enthalpy influences the solubility of a solute in an absolute manner and in its temperature dependence. The vapor pressures and the corresponding vaporization enthalpies influence the transport and persistence of organic pollutants in the environment. Sublimation enthalpy is an experimental measure of the intermolecular potentials that lead to the specific packing observed in the solid state.

EXPERIMENTAL STUDY

The vaporization enthalpies of 16 PCBs have been determined in this study using correlation gas chromatography. This study was prompted by the realization that the vaporization enthalpies previously reported by Falconer and

Bidleman for the standard compounds used in their study, octadecane and eicosane, were measured at 340 and 362 K, respectively, rather than at 298 K [1, 2]. The value used by Bidleman for the vaporization enthalpy of octadecane was 84.5 kJ/mol at 340 K; the value used for eicosane was 93.4 kJ/mol at 362 K. The vaporization enthalpies of these two standard compounds at 340 and 362 K respectively, were used to calculate the vaporization enthalpies of 32 PCBs at 298.15 K.

More recently, Ruzicka and Majer have recommended the following vaporization enthalpies for octadecane: $\Delta_{vap}H_m$ (298.15 K) = 91.4 kJ/mol and eicosane: $\Delta_{vap}H_m$ (298.15 K) = 101.8 kJ/mol [3]. A comparison of the recommended values of $\Delta_{vap}H_m$ at 298.15 K and 340 K reveals a difference of approximately 7 kJ/mol for octadecane. A difference of 8 kJ/mol was observed in the value of $\Delta_{vap}H_m$ for eicosane at 298.15 K and 362 K. In view of the large discrepancy in the recommended values at 298.15 K and the values used by Bidleman, and the fact that many of the vaporization enthalpies reported for PCBs are based on the values of two standards, we decided to re-examine the vaporization enthalpies of a representative number of PCBs. The measurements reported in this work are based on the technique of correlation gas chromatography.

Correlation gas chromatography utilizes the correlation observed between the vaporization enthalpy of several standards and their enthalpy of transfer from solution in the GC stationary phase to the gas phase. The enthalpy of transfer from solution in the GC stationary phase to vapor phase ($\Delta_{soln}{}^{vap}H_m$) is measured by gas chromatography from the temperature dependence of the retention time. The corrected retention time, t_r, for each compound is calculated from the difference between the retention times of the solute and the non-retained reference, usually the solvent. The corrected retention time measures the time each component spends on the column and it is inversely proportional to the vapor pressure of the compound on the stationary phase of the column [4]. Since the viscosity of helium increases with increasing column temperature, the retention time of the non-retained reference increases while that of the other solutes decreases [5]. This is a good indication of whether the solvent is truly unretained at the temperature of the experiment. Since the solutions are dilute, each solute presumably dissolves in the stationary phase. The values of $\Delta_{soln}{}^{vap}H_m$ can be measured for both solids and liquids.

In order to ensure that the vaporization enthalpies were not dependent on one set of standards, 18 different combinations of PCBs and standard compounds were examined under different experimental conditions. The standard compounds used in this study range from *n*-alkanes, such as octadecane, eicosane, and dodecane, to polychlorinated standards, such as *p,p'*-DDT, hexachlorobenzene, and 2,6-dichlorostyrene.

Results and Discussion

A plot of ln $(1/t_r)$ against $(1/T)$, similar to the Clausius-Clapeyron plot of ln (p) against $(1/T)$, results in a straight line whose slope is, by analogy,

$-(\Delta_{soln}{}^{vap}H_m)/R$, where R is the gas constant. All plots were characterized by correlation coefficients (r^2) greater than 0.99. The vaporization enthalpy of the target compound is calculated from the equation of the line obtained from the plot of $\Delta_{soln}{}^{vap}H_m$ against $\Delta_{vap}H_m$ (literature, 298.15 K) for the reference compounds.

A summary of the vaporization enthalpies determined from the correlation between $\Delta_{vap}H_m$ (298.15 K) and $\Delta_{soln}{}^{vap}H_m$ for each compound in each mixture is given in Table I. The results in Table I demonstrate that the values of $\Delta_{vap}H_m$ (298.15 K) calculated from the correlation equations are in good agreement with the reported literature values. A comparison of the vaporization enthalpies determined by correlation gas chromatography and the literature values is depicted in Table II. The third column in Table II represents the results from our analysis of PCBs. However, if the values of $\Delta_{vap}H_m$ (298.15 K) for octadecane and eicosane, reported by Ruzicka and Majer [3], are used to correct the vaporization enthalpies of PCBs reported by Falconer and Bidleman [2], the results are in good agreement with the values determined by our study. The last column in Table II reports these corrected values.

TABLE I. SUMMARY OF THE CORRELATION GC AND LITERATURE VALUES OF $\Delta_{vap}H_m$ (kJ/mol) AT 298.15 K FOR ALL COMPOUNDS

Compounds	No. Of Measurements	Correlation GC[a]	Literature Values[b]
1-chloronaphthalene	6	64.0 ± 0.3	64.5
1-chlorodecane	6	64.0 ± 0.2	64.0
1-chlorododecane	10	73.9 ± 1.4	75.0
1-chlorohexadecane	12	96.4 ± 0.9	98.0
2,6-dichlorostyrene	8	53.8 ± 1.5	53.0
p,p'-DDT	6	106.1 ± 1.3	106.0
hexachlorobenzene	4	74.4 ± 0.7	72.3
dodecane	3	62.1 ± 0.2	61.5
tetradecane	3	72.1 ± 0.0	71.7
octadecane	5	91.3 ± 2.9	91.4
eicosane	8	102.8 ± 2.2	101.8
tricosane	2	118.7 ± 0.0	120.5
pentacosane	4	128.6 ± 2.2	126.9
hexacosane	2	136.4 ± 0.2	139.1
biphenyl	4	64.5 ± 2.2	65.6
m-terphenyl	4	97.2 ± 0.3	95.0
anthracene	2	79.1 ± 0.0	77.9

1[c]	2	72.1 ± 2.0	-
2	2	74.3 ± 1.1	-
3	8	71.6 ± 0.7	-
7	4	75.4 ± 1.5	-
9	2	76.8 ± 0.4	-
11	2	81.0 ± 0.2	-
15	2	81.4 ± 0.3	-
18	4	80.2 ± 0.9	-
49	2	87.4 ± 0.8	-
53	4	84.9 ± 0.6	-
95	2	92.3 ± 0.6	-
96	2	89.6 ± 0.2	-
103	2	91.6 ± 0.5	-
153	2	103.5 ± 0.1	-
156	2	112.6 ± 0.4	-
171	2	109.1 ± 0.0	-

[a] The vaporization enthalpies ($\Delta_{vap}H_m$) are average values that have been calculated from the regression equation from each correlation GC experiment. The uncertainties represent the standard deviation. [b] The literature references for the vaporization enthalpies for each standard compound used in this study are given in [6]. [c] The number refers to the IUPAC designation of PCBs.

TABLE II. COMPARISON OF VAPORIZATION ENTHALPIES ($\Delta_{vap}H_m$ / kJ/mol) OF PCBS DETERMINED BY CORRELATION GC (AVERAGE VALUES) AND VALUES REPORTED IN THE LITERATURE [1,2]

IUPAC no.	Compounds	Correlation GC	Bidleman's Original Values	Bidleman's Values Corrected to 298.15 K
1	2-monochlorobiphenyl	72.1	64.45	70.00
2	3-monochlorobiphenyl	74.3	66.56	72.29
3	4-monochlorobiphenyl	71.6	66.79	72.54
7	2,4-dichlorobiphenyl	75.4	73.54	79.88
9	2,5-dichlorobiphenyl	76.8	73.95	80.31
11	3,3'-dichlorobiphenyl	81.0	75.36	81.85
15	4,4'-dichlorobiphenyl	81.4	76.03	82.58
18	2,2',5-trichlorobiphenyl	80.2	NA	-
49	2,2',4,5'-tetrachlorobiphenyl	87.4	NA	-
53	2,2',5,6'- tetrachlorobiphenyl	84.9	78.77	85.55
95	2,2',3,5',6-pentachlorobiphenyl	92.3	NA	-
96	2,2',3,6,6'- pentachlorobiphenyl	89.6	NA	-
103	2,2',4,5',6- pentachlorobiphenyl	91.6	NA	-
153	2,2',4,4',5,5'-hexachlorobiphenyl	103.5	91.43	99.30
156	2,3,3',4,4',5-hexachlorobiphenyl	112.6	94.76	102.92
171	2,2',3,3',4,4',6-heptachlorobiphenyl	109.1	95.89	104.14

Correlation gas chromatography is a fast and reliable method of evaluating vaporization enthalpies because it is based on the vaporization enthalpies of several standard compounds whose values are well established in the literature. The vaporization enthalpies of PCBs at 298.15 K in this study are considerably larger than the values reported in literature. The data from this study will influence the estimation of other important physicochemical properties of PCBs, for example, heats of partitioning. The environmental implications associated with the higher values of vaporization enthalpy of PCBs can be linked with their rate of dispersion.

COMPUTATIONAL STUDY

Quantitative Structure-Activity Relationship (QSAR) models and Comparative Molecular Field Analysis (CoMFA), a three-dimensional QSAR (3D-QSAR) paradigm, are based on the assumption that most intermolecular interactions are dependent on the molecular structure. Three-dimensional Quantitative Structure-Property Relationship (3D-QSPR) models have been constructed using CoMFA to correlate the fusion enthalpy at the melting point, sublimation enthalpy at 298.15 K, and vaporization enthalpy at 298.15 K, of a series of PCBs with their CoMFA-calculated properties. The different alignment schemes, such as *inertial*, *atom fit* and *field fit*, served as a measure of the statistical quality of the models. The CoMFA models were also developed using different partial charge formalisms, namely, electrostatic potential charges (ESP) and Gasteiger-Marsili (GM) charges.

Atom fit alignment of molecules performs a least-squares fit between two molecules by matching pairs of atoms. The quality of the fit is denoted by the root-mean-square-distance (RMSD) value computed for the matched atoms. *Field fit* alignment seeks conformations in a target molecule whose external fields most closely resemble those of a template molecule. The *as is* option in the *Database Align* module in SYBYL 6.5 molecular modeling software (Tripos, Inc., St. Louis, MO) was used to align PCBs to the biphenyl template. This alignment scheme does not change the orientation of the molecules. *Inertial* alignment of PCBs to the biphenyl template, in which only carbon atoms of the biphenyl ring were considered for alignment, was also performed in SYBYL. Partial atomic charges derived from electrostatic potentials (ESP) and Gasteiger-Marsili (GM) methods have been used in conjunction with *atom fit* alignment of molecules.

Methodology

The QSPR table was constructed with rows containing values of the thermodynamic properties for each compound in the data (training) set as the dependent variable. The columns contained the steric (van der Waals/Lennard-Jones 12-6 function) and electrostatic (Coulombic) potential energy fields as the independent variables. Partial least squares (PLS), a widely used regression technique, expresses the target property in terms of linear combinations of the independent variables. An initial PLS analysis was performed using the "leave-one-out" cross validation procedure to determine the optimum number of principal components required for the subsequent analysis of data. The optimum number of components is that which yields the largest cross-validated r^2 (r_{cv}^2) value. The PLS analysis was repeated without cross validation using the optimum number of components. This final analysis resulted in a predictive model and associated conventional r^2 values.

Results and Discussion

The CoMFA-PLS analysis of $\Delta_{fus}H_m(T_{fus})$ using *atom fit* alignment and ESP charges for a training set of 17 PCBs yields cross-validated r^2 value of 0.702 and conventional r^2 value of 0.938. The results of the CoMFA-PLS analysis of $\Delta_{fus}H_m(T_{fus})$ using *atom fit* alignment and Gasteiger-Marsili charges for the training set of 17 PCBs also demonstrate an excellent correlation between experimental and CoMFA-predicted values (Figure 1). The value of cross-validated r^2 was 0.783 and conventional r^2 was 0.955. The most predictive model for vaporization enthalpy ($\Delta_{vap}H_m$(298.15 K)), with *atom fit* alignment and Gasteiger-Marsili charges, yielded r^2 values 0.852 (cross-validated) and 0.996 (conventional) (Figure 2). Among the five different CoMFA models constructed for sublimation enthalpy, the model that combined *atom fit* alignment and ESP charges yielded the greatest self-consistency ($r^2 = 0.937$) and internal predictive ability ($r_{cv}^2 = 0.727$) (Figure 3).

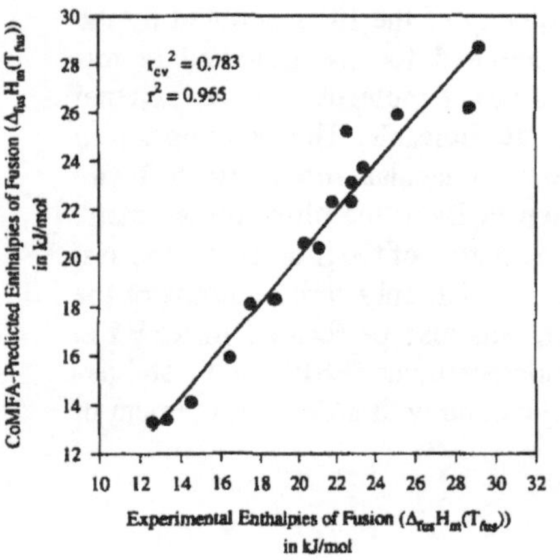

Figure 1. A plot of the CoMFA-predicted (*atom fit* alignment, Gasteiger-Marsili charges) fusion enthalpies ($\Delta_{fus}H_m(T_{fus})$) against experimental fusion enthalpies for PCBs in the training set.

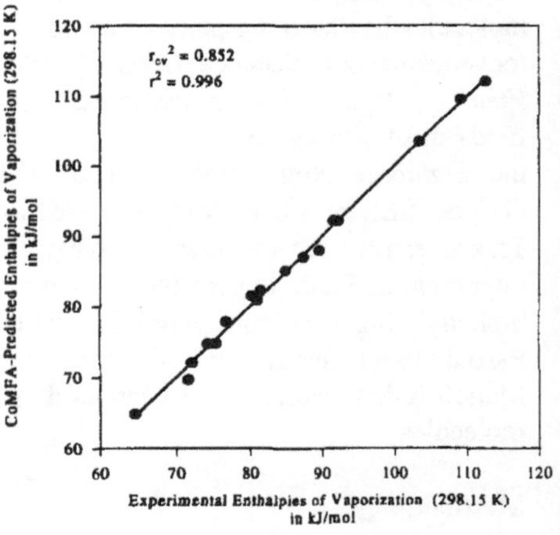

Figure 2. A plot of the CoMFA-predicted (*atom fit* alignment, Gasteiger-Marsili charges) vaporization enthalpies (($\Delta_{vap}H_m$(298.15 K)) against experimental vaporization enthalpies for PCBs in the training set.

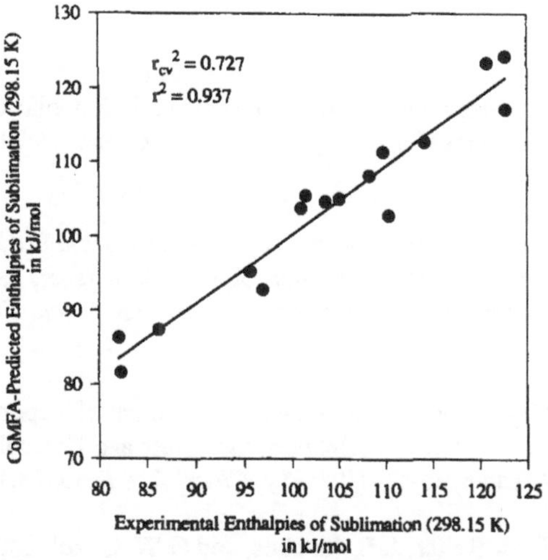

Figure 3. A plot of the CoMFA-predicted (*atom fit* alignment, electrostatic potential charges) sublimation enthalpies ($\Delta_{sub}H_m$(298.15 K)) against experimental sublimation enthalpies for PCBs in the training set.

The thermodynamic properties of PCBs increase with the number of chlorine atoms on the biphenyl rings. PCBs that are substituted at the *meta* and *para* positions are more nearly coplanar compared to *ortho*-substituted polychlorinated biphenyls. This results in stronger intermolecular interactions, and higher values of enthalpy for *m*, *p*-substituted PCBs compared to ortho-substituted PCBs. In light of the high statistical quality of the models developed in this study, the models were used to predict the thermodynamic properties of PCBs for which experimental values are not available in the literature.

The excellent predictive ability of the models derived in this study may encourage the development of similar models for other compounds of significant environmental concern, namely, polychlorinated dibenzo-p-dioxins and polychlorinated dibenzofurans. The techniques developed in this project can be used to model vapor pressures and octanol/air partition coefficients, which influence the transportation of polychlorinated biphenyls in the environment. The models may help in the analysis of properties that contribute to the bioaccumulation of organic contaminants in the environment. This knowledge can be used to design substitutes for these chemicals, which are as effective and as economical, but more degradable and less persistent. This environmental issue presents a considerable scientific and industrial challenge.

This research is partially funded by a grant to WJW from the U.S. Environmental Protection Agency's Science to Achieve Results (STAR) program.

REFERENCES

1. Bidleman, T.F. 1984. "Estimation of Vapor Pressures for Nonpolar Organic Compounds by Capillary Gas Chromatography," *Anal. Chem.*, 56:2490-2496.

2. Falconer, R.L. and T. F. Bidleman. 1994. "Vapor Pressures and Predicted Particle/Gas Distributions of Polychlorinated Biphenyl Congeners as Functions of Temperature and Ortho-Chlorine Substitution," *Atm. Env.*, 28: 547-554.

3. Ruzicka, K. and V. Majer. 1994. "Simultaneous Treatment of Vapor Pressures and Related Thermal Data Between the Triple and Normal Boiling Temperatures for *n*-Alkanes C5-C20," *J. Phy. Chem. Ref. Data*, 23:1-39.

4. Nichols, G., J. Orf, S. M. Reiter, J. S. Chickos, and G.W. Gokel. 2000. "The Vaporization Enthalpies of Some Crown Polyethers by Correlation Gas Chromatography," *Thermochim. Acta*, 346:15-28.

5. Chickos, J., S. Hosseini, and D.G. Hesse. 1995. "Determination of Vaporization Enthalpies of Simple Organic Molecules By Correlations of Changes in Gas Chromatographic Retention Times," *Thermochim. Acta*, 249: 41-62.

6. Puri, S., J. S. Chickos, and W. J. Welsh. 2001. "Determination of Vaporization Enthalpies of Polychlorinated Biphenyls by Correlation Gas Chromatography," *Anal. Chem.*, 73(7):1480-1484.

Atmospheric Chemistry of Alternative Fuels

Joseph S. Francisco

Department of Chemistry and Department of Earth and Atmospheric Sciences

Purdue University

West Lafayette, Indiana 47907

Abstract

At the present time, fuels derived from petroleum supply virtually all of the world's transportation needs. The combustion of petroleum based fuels leads to the formation of particulates (soot), NO_x, SO_x, ozone, and ultimately to the formation of air pollution. The 1990 Clean Air Act mandated reformulated gasoline programs. The oxygenated fuels program requires the use of fuels containing 2.7% oxygen by weight in the wintertime to reduce CO emissions and 2.0% oxygen by weight in the summer to reduze ozone formation. In this talk, the environmental acceptability of alternative fuels will be presented. Experimental and computational chemistry results will be presented to highlight key mechanistic steps in the atmospheric oxidation of alternative fuels.

Computing the properties of the Cu-thioarsenite complex

J. A. Tossell
Dept. of Chemistry and Biochemistry
University of Maryland
College Park, MD 20742
U.S.A.
Tossell@chem.umd.edu

In aqueous solutions in equilibrium with the mineral assemblage CuS(covellite-$Cu_{1.8}S$(digenite)-Cu_3AsS_4(enargite), a Cu-thioarsenite species, CuAsS(SH)(OH) is observed to make a major contribution to both Cu and As solubility (Clarke and Helz, Environ. Sci. Technol., 2000, 34, 1483-1488), confirming its high stability and establishing the presence of a novel direct bond between Cu (formally Cu(I)) and As (formally As(III)). To provide further evidence for the existence of this complex it would be desirable to concentrate it (perhaps in a nonpolar solvent, since it is neutral and not extremely polar) and to measure its spectral properties. To assist in the confirmation of its existence we have calculated a number of different properties for this complex at high quantum mechanical levels, including its vibrational, visible-UV, X-ray Absorption Near Edge and Cu and As NMR spectra. The visible-UV spectrum has been calculated using both Hartree-Fock and density-functional-theory based methods and the different methods have been tested against the known properties of the gas-phase CuCl molecule. Mulliken population analyses, Natural Bond Orbital analyses and contour plots of molecular orbital electron densities are also used to characterize the bonding with the Cu-thioarsenite complex. We have also calculated the hydration energy of this complex in both water and nonpolar solvents using polarized dielectric continuum methods.

Holly A. Bamford
Chesapeake Biological Laboratory
University of Maryland
P.O. Box 38, 1 Williams Street
Solomons, MD 20688-0038
T 410-326-7204
F 410-326-7341
bamford@cbl.umces.edu

Presenter: H.A. Bamford

The Prediction of Henry's Law Constants and its Temperature Dependence for Polychlorinated Biphenyls from Computational and Experimental Data. Bamford, H.A.[1], Poster, D.L.[2], Allison, T.C. [2], Baker, J.E.[1], [1] Chesapeake Biological Laboratory, Solomons, MD, [2] National Institute of Standards and Technology, Gaithersburg, MD. The Henry's law constants for twenty six polychlorinated biphenyls (PCBs) were experimentally determined between 4 °C and 31 °C using a gas-stripping apparatus. The results of these experimental determinations were then compared with values derived from a parameterized continuum dielectric solvation model (SM5.42) at the PM3 level of theory. Furthermore, the temperature dependence of each PCB was also modeled using the van't Hoff equation to calculate the enthalpy (ΔH) and entropy (ΔS) of phase change. A plot of ΔS *versus* ΔH for the 26 congeners showed a linear correlation which suggests they are inherently related. This relationship, the enthalpy-entropy compensation effect, was used to estimate the Henry's Law constants for the remaining PCB congeners as functions of temperature. These results, along with a comparison of the experimental and theoretical determinations will be discussed.

SUCCESS OF DENSITY FUNCTIONAL THEORY IN PREDICTING ENTHALPY CHANGES ASSOCIATED WITH TWO-ELECTRON REDUCTION OF POLYCHLORINATED BIPHENYLS

LISA A. TOTTEN
Department of Environmental Sciences
14 College Farm Road
Rutgers University
New Brunswick, NJ 08901, U.S.A.

XUEFENG LIU
Department of Earth and Environmental Engineering
Columbia University, New York, NY

DAVID J. BRAUN
NADA ASSAF-ANID
Chemical Engineering Department
Manhattan College
Manhattan College Parkway
3825 Corlear Avenue
Riverdale, New York City, NY 10471-4098, U.S.A.

INTRODUCTION

Accurate reduction potentials for polychlorinated biphenyls (PCBs) and other aromatic halides would be useful indicators of their ability to undergo reductive dehalogenation and would allow the construction of quantitative structure-activity relationships (QSARs) to predict the kinetics of the reaction. Two-electron reduction potentials (E_2 values) describing the hydrogenolysis of PCBs help in the assessment of the thermodynamic feasibility of the reaction. If reduction is thought to occur via single electron transfer, Marcus theory suggests that one-electron reduction potentials (E_1 values) would be useful for the prediction of reaction kinetics. Unfortunately, E_1 and E_2 values for PCBs are difficult to obtain and published values are of questionable accuracy. This research aims to identify methods for predicting reduction potentials for PCBs via computational chemistry and thermodynamic calculations.

For C_1 and C_2 halocarbons, one-electron reduction is thought to occur via a dissociative mechanism in which cleavage of the carbon-halogen bond occurs simultaneously with transfer of a single electron *[1]*:

$$RX(aq) + e^- \rightarrow R^\bullet(aq) + X^-(aq) \qquad (1)$$

This reaction results in a carbon-centered radical and is irreversible under normal conditions. E_1 for alkyl and vinyl halides have been calculated from thermodynamic data *[2, 3]*, measured via electrochemical techniques *[4]*, or estimated from computational chemistry *[5, 6]*. In many cases, the C-X bond dissociation energy (which is closely related to E_1) is estimated from computational chemistry *[7, 8]*.

In contrast, one-electron reduction of aromatic halides such as PCBs has typically been observed to occur via a non-dissociative mechanism, resulting in the formation of the anion radical ($RX^{\bullet-}$) *[9]*, which rapidly undergoes R-X bond cleavage to yield a carbon centered radical and a halide ion. Although the initial ET step is reversible, because R-X bond cleavage is fast, the reaction is macroscopically irreversible. Because the heats of formation of aromatic radical anions are unavailable, E_1 values cannot be calculated from thermodynamic data for PCB and other aromatic halides. E_1 values for PCBs may be approximated by cyclic voltammetry studies and half-wave potentials, but the macroscopic irreversibility of the reaction complicates these measurements. In addition, this data is only available for a very limited number of congeners *[10, 11]*, and the experiments are typically conducted in solvents other than water, requiring that the measured reduction potential be corrected for the change in solvent. E_1 values can be predicted via computational chemistry by estimating the energies of PCB congeners and their radical anions. The difference in solvation energy between the neutral parent and the negatively charged product is large, however, and cannot be estimated from the Henry's Law constant of the anion, because this parameter cannot be measured. This makes the derivation of E_1 values for aromatic halides from computational chemistry problematic. We are currently pursuing several strategies to solve this problem.

In contrast to E_1, derivation of E_2 values for aromatic halides should be relatively straightforward. E_2 values are typically calculated from published thermodynamic data. For example, Vogel *et al.* *[12]* and Totten and Roberts *[3]* have used published enthalpy and entropy data to calculate E_2 values for halogenated methanes, ethanes and ethylenes. The resulting calculated reduction potentials are reasonably accurate because a large amount of experimentally determined thermodynamic data for these compounds are available. In contrast, thermodynamic properties of most PCBs and aromatic halides have not been measured experimentally, although estimates of such properties for PCB, chlorobenzenes, and other aromatic chlorides derived from bond-contribution methods have been published *[13, 14]*. The bond-contribution factors for aromatic halides tend to be derived from a very small database of experimentally determined thermodynamic data for this class of compounds, suggesting that the factors may not be highly accurate *[15]*. In addition, it has been noted that such bond-contribution factors are not applicable to aliphatic polyhalides, because the effects of multiple halogens are not strictly additive *[15, 16]*. Thus it is not clear whether bond contribution methods are capable of accurately predicting thermodynamic properties of polychlorinated aromatics.

Given the possible shortcomings of the bond-contribution method of estimating thermodynamic properties of aromatic chlorides, computational chemistry offers a promising method of deriving accurate E_2 values for these compounds. Computational models are capable of generating values for the enthalpy and entropy of formation of a compound, as well as its solvation energy (via the use of various solvation models). Derivation of entropy terms requires that a frequency calculation be performed. We estimate that the entropy and solvation terms contribute only about 15% and 6% respectively to the overall free energy change of the hydrogenolysis reaction (described in more detail below). Frequency calculations and solvation modeling both consume large amounts of computing time. Future research will incorporate these calculations into the computational framework, but for the present study, they were not performed.

In this work, the ability of computational chemistry to accurately describe the enthalpy changes associated with hydrogenolysis of aromatic chlorides is investigated. The goals of this work are to generate computational estimates of enthalpy changes associated with hydrogenolysis (ΔH_2) of aromatic chlorides, a parameter which is closely related to E_2, and to evaluate the accuracy of these values by critically comparing them with ΔH_2 and E_2 values derived from thermodynamic data from bond contribution estimates and experimental determinations (where available), and with expected trends in reduction potentials predicted from Hammett theory.

METHODS

Becke's 3-parameter hybrid functional using the Lee-Yang-Parr gradient-corrected correlation functional (B3LYP) was employed in this work *[17]*. This method was chosen because it has been shown to result in the lowest average absolute deviation among several DFT methods evaluated in predicting $\Delta H_f°$ for molecules the G2-2 data set *[18]*. The 6-31G* basis set was chosen as a starting point for calculations. The success of these calculations will determined whether a larger basis set may be necessary to adequately model aromatic chlorides.

The enthalpy of the following reaction was derived for each congener from the computational energies of the products and reactants:

$$RCl\ (g) + H^{\bullet}\ (g) \rightarrow RH\ (g) + Cl^{\bullet}\ (g) \qquad (2)$$

ΔH_2 for this reaction was also calculated from published thermodynamic data in order to make valid comparisons with the computationally derived ΔH_2. This reaction enthalpy is similar to the enthalpy of the hydrogenolysis reaction because the enthalpy of the following reaction is constant:

$$H^{+}\ (aq) + Cl^{\bullet}\ (g) \rightarrow H^{\bullet}\ (g) + Cl^{-}\ (aq) \qquad (3)$$

Thermodynamic data for PCBs were taken from Holmes *et al.* [13], who used vapor pressure and solubility data for PCBs from Burkhard *et al.* [19] to convert gas-phase free energies of formation into aqueous values. More recently, Falconer and Bidleman [20] published a comprehensive tabulation of vapor pressures of PCB congeners which are often cited as the most accurate values currently available. As yet there is no generally accepted database of solubilities of PCBs, especially for the high MW congeners. An alternative method of converting gas phase to aqueous free energies would be to use Henry's Law constants (HLC). Published values of HLC for individual PCB congeners can range over an order of magnitude or more, however [19, 21, 22]. We estimate that the contribution of the change in solvation between the reactant and product contributes only at most 6 kJ mol^{-1}, or less than 10%, of the overall free energy change of the reaction. Thus we used the thermodynamic data of Holmes *et al.* [13] without revision. This thermodynamic data was also used to calculate E_2 values of PCBs and chlorobenzenes. The framework for performing calculations of E_2 at pH 7 and 1 mM chloride is described in detail in Vogel *et al.* [12] and Totten and Roberts [3].

RESULTS

E_2 values for PCBs calculated from published thermodynamic data range from 0.46 to 0.64 V vs. SHE [23-25]. For comparison, Dolfing and Harrison [14] calculated E_2 values for chlorobenzenes from bond-contribution thermodynamic data ranging from 0.31 V for chlorobenzene to 0.48 V for hexachlorobenzene. Although it might be expected that the lowest values of E_2 would result from hydrogenolysis of the monochlorobiphenyls, in fact the lowest value is for removal of an ortho chlorine from 2,2',3,4,4',5,5'-heptachlorobiphenyl to form 2,3,3',4,4',5-hexachlorobiphenyl. The highest calculate E_2 value is for removal of a meta chlorine from 2,2',3,3',4,4',5,5'-octachlorobiphenyl. Removal of a chlorine in the ortho position might be expected to be more thermodynamically favorable than removal of a meta or para chlorine. It is true that the calculated E_2 value for reduction of 2-chlorobiphenyl is greater than that for 3- or 4-chlorobiphenyl, but in general there is no statistical difference between reduction potentials calculated from the bond-contribution thermodynamic data for removal of ortho and meta or para chlorines.

ΔG_{rxn} determined for PCBs at standard conditions (pH 0, 1 M chloride) ranges from -100 to -134 kJ mol^{-1}. The change in entropy from reactant PCB to product PCB ranges from 15.3 to 0.8 kJ mol^{-1}. Changes in solvation were found to range from 5.9 to 0.4 kJ mol^{-1}. Both parameters thus constitute only a small fraction of the overall free energy change of the hydrogenolysis reaction, further justifying the decision not to perform frequency and solvation calculations in the computational framework.

ΔH_2 values (for reaction 2) of PCBs range from -104 to -143 kJ mol^{-1} via the computational method and -58 to -86 kJ mol^{-1} from the bond-contribution thermodynamic data of Holmes *et al.* *[13]*. Thus while the bond contribution method predicts that ΔH_2 values range over only 28 kJ mol^{-1} for the PCBs (corresponding to a range in E_2 of 0.15 V), the computational method predicts that ΔH_2, and therefore E_2, will vary over a broader range (39 kJ mol^{-1} or 0.20 V). The bond contribution data appears to systematically overestimate ΔH_2 for loss of ortho chlorines relative to the computational values (Figure 1). Also, the computational method underpredicts ΔH_2 by 18-37 kJ mol^{-1}, relative to the bond contribution values. A striking feature of Figure 1 is the lumping of the thermodynamic ΔH_2 values into three distinct groups. This is probably an artifact of the way in which the bond-contribution method estimates $\Delta H_f°$ for

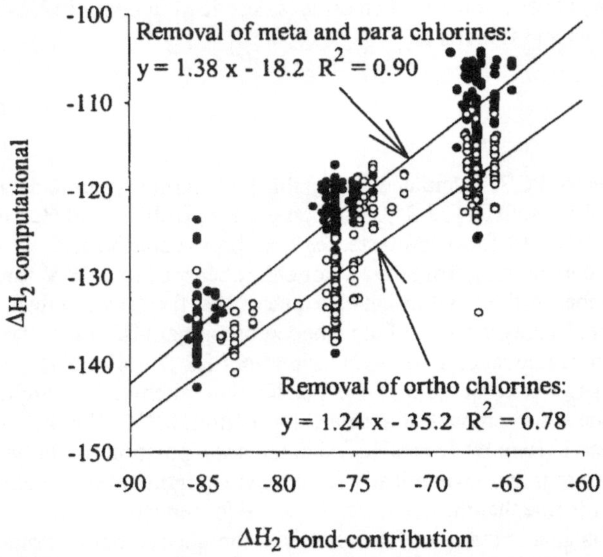

Figure 1: Enthalpy of the hydrogenolysis reaction (ΔH_2) from computational methods plotted versus the same parameter calculated from bond-contribution estimations of the heats of formation of PCB congeners.

PCBs.

Because there is considerable uncertainty associated with the bond-contribution estimates of $\Delta H_f°$, and therefore of ΔH_2 value calculated using this data, it is important to examine the relationship between the computational ΔH_2 values and experimental data. Because very little experimental thermodynamic data is available for PCBs, this comparison must utilize data for chlorobenzenes and other alkyl halides. $\Delta H_f°$ values for all of the chlorobenzenes determined by Platonov *et al. [26-28]* were used in the present calculations. Other ΔH_2 values were calculated from the $\Delta H_f°$ values recommended by Totten and Roberts *[3]*. Figure 2 demonstrates that the computational method is capable of accurately predicting *relative* ΔH_2 values for chlorinated methanes and ethanes, as well as many chlorobenzenes, although the slope of this relation is not one and the intercept is not zero. All calculations involving trichlorobenzenes (i.e. hydrogenolysis of tetra- and trichlorobenzenes) deviate significantly from this relationship. Because it seems unlikely that the computational method would accurately describe all of the other chlorobenzenes but not the trichloro homologues, we suspect that the experimentally determined $\Delta H_f°$ values for these compounds may be in error. Table I lists alternate values of $\Delta H_f°$ for the trichlorobenzenes from Yan *et al. [29]*, and bond-contribution estimates of $\Delta H_f°$ from Domalski and Hearing *[15]*. Use of these alternate

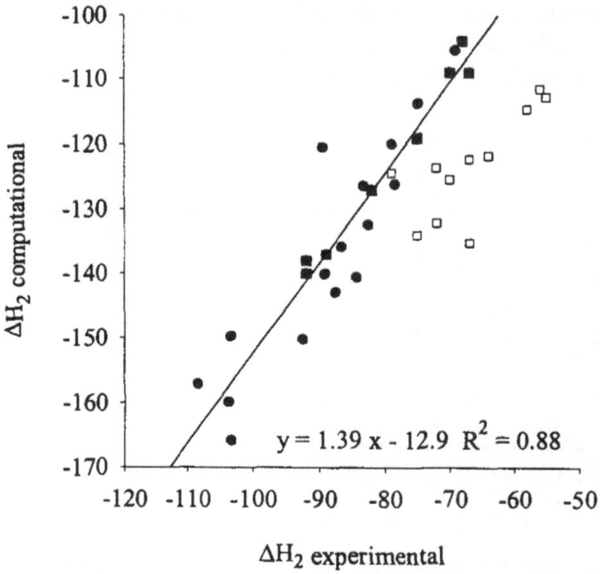

Figure 2: ΔH_2 from computational calculations versus experimental determinations for halogenated methanes and ethanes (circles), most chlorobenzenes (filled squares), and reactions involving trichlorobenzenes (open squares).

Table I: Heats of formation ($\Delta H_f°$) of trichlorobenzene isomers (in kJ mol^{-1}).			
isomer	**1,2,3**	**1,2,4**	**1,3,5**
Platonov et al.	3.78	-8.05	-13.36
Yan, et al.	8.23	4.85	-2.57
Domalski and Hearing	4.34	-5.16	-24.66

$\Delta H_f°$ values results in ΔH_2 values that do not significantly improve the correlation seen in Figure 2. Thus the $\Delta H_f°$ values for trichlorobenzenes are not well established. If they are indeed in error, Figure 2 suggests correction factors (slope and intercept) could be used to generate accurate absolute values of ΔH_2 from computational estimates of ΔH_2 for PCBs and other chlorinated aromatics.

Another method by which the success of the computational method at predicting ΔH_2 might be judged is to examine predictions based on Hammett theory. ΔH_2 for chlorobenzenes is plotted versus the sum of Hammett constants ($\Sigma\sigma$) in Figure 3. Because $\sigma_{ortho,Cl}$ is not available for chlorobenzenes, this value was generated via a minimization of squared residuals regression for each data set and was found to equal 2.01, 1.23, and 1.17 for the experimental, bond-contribution, and computational ΔH_2 values, respectively. These values are high in comparison to $\sigma_{meta,Cl}$ (0.37) and $\sigma_{para,Cl}$ (0.23), and also as compared to $\sigma_{ortho,Cl}$ for phenols (0.68) and anilines (0.67) *[1]*. The correlation with Hammett constants is best for the computational ΔH_2 values and significantly worse for the bond-contribution values. The correlation is surprisingly poor for

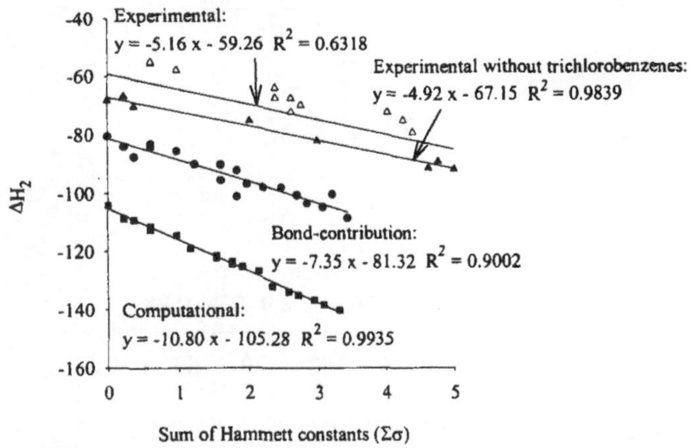

Figure 3: ΔH_2 for chlorobenzenes plotted versus the sum of Hammett constants ($\Sigma\sigma$). ΔH_2 values are derived from computational calculations (squares), bond-contribution thermodynamic data (circles) and experimental data (triangles). The correlation of the experimental ΔH_2 values with $\Sigma\sigma$ improves dramatically if the ΔH_2 values involving trichlorobenzenes (open triangles) are excluded from the correlation.

the experimental ΔH_2 values. Again this may be due to inaccurate values for $\Delta H_f°$ of trichlorobenzenes. If these points are removed from the regression, the correlation improves greatly.

Taken together, these results suggest that the B3LYP/6-31G* computational method (a relatively low level of computational theory) may be used to generate accurate values ΔH_2 via the use of correction factors resulting from the correlation between computational and experimental thermodynamic ΔH_2 values. The computational ΔH_2 values would appear to be more accurate that those generated from bond-contribution estimates of $\Delta H_f°$ for aromatic halides, and may in fact be more accurate than those generated from the limited amount of experimental thermodynamic data available for this class of compounds.

ΔH_2, the enthalpy change associated with hydrogenolysis, is the most important parameter in the calculation of E_2 values. Thus this computational method should be capable of generating accurate and useful two-electron reduction potentials for the hydrogenolysis of aromatic chlorides. Further research will focus on performing a limited number of frequency calculations on PCBs and chlorobenzenes to determine the entropy changes associated with hydrogenolysis, and on characterizing the solvation changes that occur during one-electron reduction of chlorinated aromatics, so that accurate one-electron reduction potentials may be determined for these compounds.

REFERENCES

1. Schwarzenbach, R. P.; Gschwend, P. M.; Imboden, D. M. *1993*. *Environmental Organic Chemistry*; Wiley and Sons: New York, .

2. Curtis, G. P. *1991*. ""Reductive Dehalogenation of Hexachloroethane and Carbon Tetrachloride by Aquifer Sand and Humic Acid"." PhD Dissertation, Stanford University.

3. Totten, L. A.; Roberts, A. L. *2001*. "Calculated one- and two-electron reduction potentials and related molecular descriptors for reduction of alkyl and vinyl halides in water," *Crit. Rev. Environ. Sci. Technol.*,

4. Mann, C. K.; Barnes, K. K. *1970*. "Reactions of the Carbon-Halogen Bond," In *Electrochemical Reactions in Nonaqueous Systems*, Marcel Dekker: New York, , *201-244*.

5. Perlinger, J. *1994*. "Reduction of Polyhalogenated Alkanes by Electron Transfer Mediators in Aqueous Solution." PhD Dissertation, Swiss Federal Institute of Technology.

6. Totten, L. A. *1999*. ""The Use Of Model And Probe Compounds To Investigate The Mechanisms Of Reductive Dehalogenation Reactions"." PhD Dissertation, The Johns Hopkins University.

7. Liu, Z.; Betterton, E. A.; Arnold, R. G. *2000.* "Electrolytic reduction of low molecular weight chlorinated aliphatic compounds: Structural and thermodynamic effects on process kinetics," *Environ. Sci. Technol., 34, 804-811.*

8. Perlinger, J. A.; Venkatapathy, R.; Harrison, J. F. *2000.* "Linear Free Energy Relationships for Polyhalogenated Alkane Transformation by Electron-Transfer Mediators in Model Aqueous Systems," *J. Phys. Chem. A, 104, 2752 - 2763.*

9. Savéant, J.-M. *1993.* "Electron transfer, bond breaking, and bond formation," *Acc. Chem. Res., 26, 455-461.*

10. Rusling, J. F.; Miaw, C. L. *1989.* "Kinetic estimation of standard reduction potentials for polyhalogenated biphenyls," *Environ. Sci. Technol., 23, 476-479.*

11. Huang, Q.; Rusling, J. F. *1995.* "Formal reduction potentials and redox chemistry of polyhalogenated biphenyls in a bicontinuous microemulsion," *Environ. Sci. Technol., 29, 98-103.*

12. Vogel, T. M.; Criddle, C. S.; McCarty, P. L. *1987.* "Transformations of halogenated aliphatic compounds," *Environ. Sci. Technol., 21, 722-736.*

13. Holmes, D. A.; Harrison, B. K.; Dolfing, J. *1993.* "Estimation of Gibbs free energies of formation for polychlorinated biphenyls," *Environ. Sci. Technol., 27, 725-731.*

14. Dolfing, J.; Harrison, B. K. *1992.* "Gibbs free energy of formation of halogenated aromatic compounds and their potential role as electron acceptors in anaerobic environments.," *Environ. Sci. Technol., 26, 2213-2218.*

15. Domalski, E. S.; Hearing, E. D. *1993.* "Estimation of the thermodynamic properties of C-H-N-O-S-halogen compounds at 298.15°K," *J. Phys. Chem. Ref. Data, 22, 805-1101.*

16. Chen, C.-J.; Wong, D.; Bozzelli, J. W. *1998.* "Standard chemical thermodynamic properties of multichloro alkanes and alkenes: a modified group additivity scheme.," *J. Phys. Chem. A, 102, 4551-4558.*

17. Foresman, J. B.; Frisch, A. *1996. Exploring Chemistry with Electronic Structure Methods;* 2nd edition ed.; Gaussian, Inc.: Pittsburgh, PA, .

18. Curtiss, L. A.; Raghavachari, K.; Redfern, P. C.; Pople, J. A. *1997.* "Assessment of Gaussian-2 and density functional theories for the computation of enthalpies of formation.," *J. Chem. Phys., 106, 1063-1079.*

19. Burkhard, L. P.; Armstrong, D. E.; Andren, A. W. *1985.* "Henry's Law constants for the polychlorinated biphenyls.," *Environ. Sci. Technol., 19, 590-596.*

20. Falconer, R. L.; Bidleman, T. F. *1994.* "Vapor pressures and predicted particle/gas distributions of polychlorinated biphenyl congeners as a function of temperature and ortho-chlorine substitution.," *Atmos. Env., 28, 547-554.*

21. Brunner, S.; Hornung, E.; Santl, H.; Wolff, E.; Piringer, O. G.; Altschuh, J.; Brüggemann, R. *1990.* "Henry's law constants for polychlorinated biphenyls: Experimental determination and structure-property relationships," *Environ. Sci. Technol., 24, 1751-1754.*

22. Bamford, H. A.; Poster, D. L.; Baker, J. E. *2000.* "Henry's Law constants of polychlorinated biphenyl congeners and their variation with temperature," *J. Chem. Eng. Data, 45, 1069-1074.*

23. Assaf-Anid, N.; Anid, P. J.; Nies, L.; Vogel, T. Estimation of the Redox Potentials and Gibbs Free Energies of Dechlorination of Polychlorinated Biphenyls. In *Society of Environmental Toxicology and Chemistry National Meeting*, Philadelphia , PA; 1999.

24. Assaf-Anid, N.; Totten, L. A.; Braun, D. Computational chemistry calculations of thermodynamic descriptors for chlorinated aliphatic compounds and PCBs. In *Society of Environmental Toxicology and Chemistry National Meeting*, Nashville , TN; 2000.

25. Totten, L. A.; Liu, X.; Braun, D. J.; Assaf-Anid, N. *2001.* "Use of computational chemistry to predict reduction potentials of polychlorinated biphenyls," *Abstr. Pap. Am. Chem. S., 41, 86-89.*

26. Platonov, V. A.; Simulin, Y. N. *1983.* "Standard enthalpies of formation of 1,2,3-trichlorobenzene, 1,2,4,5-tetrachlorobenzene, and hexachlorobenzene," *Russ. J. Phys. Chem. (Engl. Transl.), 57, 840-842.*

27. Platonov, V. A.; Simulin, Y. N. *1985.* "Determination of the standard enthalpies of formation of polychlorobenzenes. III. The standard enthalpies of formation of mono-1,2,4- and 1,3,5-tri-, and 1,2,3,4- and 1,2,3,5-tetrachlorobenzenes," *Russ. J. Phys. Chem. (Engl. Transl.), 59, 179-181.*

28. Platonov, V. A.; Simulin, Y. N.; Rozenberg, M. M. *1985.* "Standard heat of formation of pentachlorobenzene. Calculation of ΔHc°(g) and ΔEs°(g) of polychlorobenzenes by the Tatevskii method," *Russ. J. Phys. Chem. (Engl. Transl.), 59, 814-817.*

29. Yan, H.; Gu, J.; An, X.; Hu, R. *1987.* "Standard enthalpies of formation and enthlpies of isomerization of trichlorobenzenes," *Huaxue Xuebao, 45, 1184-1187.*

COMPUTATIONAL MODELS FOR PREDICTING THE THERMODYNAMIC PROPERTIES OF POLYCHLORINATED BIPHENYLS (PCBs)

Swati Puri, William J. Welsh* and James S. Chickos
Department of Chemistry and Biochemistry, University of Missouri–St. Louis
8001 Natural Bridge Road, St. Louis, Missouri 63121
*Corresponding Author: Email Address: wwelsh@umsl.edu

Polychlorinated biphenyls (PCBs) are acknowledged to possess many attractive properties, namely non-flammability and chemical stability, which contributed to their widespread use as components of pesticides, plasticizers and adhesives. Unfortunately, these same features have contributed to their classification as organic pollutants. A study of the thermodynamic properties of PCBs is essential to assess the impact of these compounds on the environment.

Simple and predictive models have been developed for the enthalpies of fusion, enthalpies of vaporization, and enthalpies of sublimation of PCBs using the three-dimensional quantitative structure-property relationship (3D-QSPR) method of comparative molecular field analysis (CoMFA). The present CoMFA models can be used to predict thermodynamic properties of PCBs for which experimentally measured values are unavailable. Various alignment schemes and formalisms for assigning partial atomic charges were evaluated to select the most statistically predictive model. The exceptionally high self-consistency ($r^2 > 0.9$) and internal predictive ability ($r_{cv}^2 > 0.5$) of the models derived in this study may encourage the development of similar models for other compounds of significant environmental concern, such as polychlorinated dibenzo-p-dioxins and polychlorinated dibenzofurans. These models may also help in the analysis of properties that contribute to the bioaccumulation of organic contaminants in the environment.

This research is partially funded by a grant to WJW from the U.S. Environmental Protection Agency's Science to Achieve Success (STAR) program.

SYMPOSIUM 8:

Tracing and Geological Engineering

A Modified Solution for Contaminant Removal by Pneumatic Fracturing

Yuan Ding
Department of Civil and Environmental Engineering
New Jersey Institute of Technology, Newark, NJ 07102

In subsurface remediation of a contaminated site, the vadose zone, as a transmittal zone tween the ground surface and groundwater, plays a critical role. Among the various hazardous astes, it is generally recognized that volatile organic compounds (VOCs) are the major ncerns. Soil vapor extraction (SVE) has been proven to be an effective technology in removing OCs from the vadose zone of geologic formations having high air permeability. From the field perience, it has been found that the SVE becomes ineffective in formations of low air rmeability such as silt and clay. The pneumatic fracturing process is an innovative technology at enhances the air permeability of the formation. This process has been recognized as a cost-fective approach in site remediation engineering.

In practice, pneumatic fracturing uses compressed clean air to create new fractures in soil rmations and to widen existing fractures in sedimentary rock formations, thereby increasing e air permeability of these formations by several orders of magnitude. The mode of ntaminant transport in the subsurface is thus changed from primarily diffusion to convection. ne efficiency of VOC contaminant removal is therefore improved tremendously.

This pneumatic fracturing enhanced the SVE process was simulated using a pair of coupled rtial differential equations. A lengthy analytical solution to the equations was obtained, which as verified with contaminant removal data from a field project (Ding, et al., 1999). In order to ply this solution in the field conveniently, a user-friendly format is desirable. This paper esents the simplification analysis and the associated conditions. The simplified solution can be adily used by engineers or decision makers directly with a regular calculator.

Corresponding author: Yuan Ding
Telephone: 973-642-7046

Email: ding@njit.edu

eference

ing, Y., Schuring, J. R., and Chan, P. C. "Engineering Solution for Predicting Contaminant Removal by Pneumatic Fracturing". ASCE Practice Periodical of Hazardous, Toxic, and Radioactive Waste Management, Vol. 3, No. 3, pp 101-106, 1999.

ing, Y., Schuring, J. R., and Chan, P. C. "Volatile Contaminant Extraction Enhanced by Pneumatic Fracturing". ASCE Practice Periodical of Hazardous, Toxic, and Radioactive Waste Management, Vol. 3, No. 2, pp 69-76, 1999.

Development of an Automated Monitoring System for a Created Pond At Great Kills Park, Staten Island, New York

Emily Ayers, Columbia University
Roelof Versteeg, Columbia University

Wetland construction and remediation is an inexact science. One of the reasons for this is that the temporal and spatial variability of a wetland, and its causes and effects are poorly known and understood. In order to enhance this understanding we need both better measurement and better models. The challenge with doing measurements on a highly variable system is that we do not know where to measure. In order to address this problem a flexible, automated monitoring system has been custom designed developed for a newly created pond in the Greatkills area of Gateway National Park
The system (comprising of a suite of sensors focused on measuring relevant parameters, controlled and sampled by a custom designed acquisiton interface written in Labview 6i) is designed to operate continuously, providing baseline data on various aspects of the pond water quality, including temperature, pH and dissolved oxygen.

These parameters will be tracked continuously throughout the self-organization of the pond. These data will allow us to correlate changes in water quality with processes at work during the development of the pond ecosystem.

Data accessibility is achieved through broadcast of the data to a geospatial database on a remote webserver running ArcIMS which allows for web based data visualization and analysis.

Design and Construction of a Coupled Subsurface Imaging/ Remediation System

Eric Stabinski, Grant Sprick and Roelof Versteeg
Columbia University, New York

Current implementations of subsurface contamination remediation systems are not satisfactory both from a cost and effectiveness point of view. The main cause underlying the problem with subsurface remediation systems is the fact that these systems can not be targeted to address the contaminants directly. This inability to target contaminants is related to the fact that there is no direct and real time information about the effect of remediation.

By integrating real time subsurface imaging capability with remediation system we could in theory significantly enhance the effectiveness of these systems. The design of such an integrated system should incorporate both the different components of the system as well as their interaction. A prototype coupled subsurface imaging and remediation system was designed and constructed and is currently operational. The results from this system demonstrate that it is feasible to provide near real time information on the changes in subsurface physical properties which are related to remediative activities. Extension of this system to a field scale system should be feasible.

Title: Bio-Optical Modeling of the Nearshore Waters Using Remote Sensing and In-Situ Measurements

Author: Sima Bagheri, Ph.D.
Affiliation: Department of Civil & Environmental Engineering
New Jersey Institute of Technology, Newark, NJ 07102

This paper reports on the development of remote sensing strategies which address problems of water quality (i.e., harmful algal blooms and their long term consequences of ecosystem changes) in New Jersey nearshore waters. The study area is the Hudson/Raritan Estuary of New York-New Jersey where there are major conflicting interests in the use of these waters. Remote sensing methods offer unique advantages for studying of recurrent hydrological phenomena on regional and local scales. The airborne/satellite sensor records the color of natural water as spectral reflectance, which is determined by changes in the spectral composition of the underwater light field (Kirk, 1984). The greater the amount and specifity of color information available, the better a remote hydrological application will generally perform. This is particularly true for estuarine applications, where independent variations of optically important water column constituents, bottom cover types and water depths all tend to occur simultaneously.

Based on Remote Sensing and in-Situ measurements optical water quality models are constructed linking the inherent optical properties (IOP) to the subsurface irradiance reflectance (R(O-)). R(O-) is a key parameter to derive to atmospherically correct the remote sensing data. Subsequently, a suite of analysis are being used to retrieve the water quality parameters (i.e., phytoplankton, suspended solids and colored dissolved organic matter) from the remote sensing signals.

Sima Bagheri

Department of Civil & Environmental Engineering
New Jersey Institute of Technology
Newark, NJ 07102
Phone: (973)596-2470
Fax: (973)596-5790
http://klimt.njit.edu/~sima/

SYMPOSIUM 9:

Thermal and Kinetic Advances in Pollution Control and Prevention

CORRELATING LAMINAR BURNING VELOCITIES USING PERFECTLY STIRRED REACTOR THEORY FOR HAZARDOUS WASTE INCINERATION

ROBERT B. BARAT

Department of Chemical Engineering
New Jersey Institute of Technology
Newark, NJ 07102

Abstract

The laminar burning velocity (LBV) of a gaseous fuel/oxidant feed is a strong function of the feed mixture composition. In this paper, published LBV values are successfully correlated with the square root of the ratio of the thermal diffusivity of the fuel/oxidant feed mixture to the space time at extinction of a hypothetical perfectly stirred reactor (PSR) burning this feed. The PSR simulations use detailed reaction mechanisms. The correlation would be especially useful for estimating the effect chemical wastes would have on the stability of incinerator flames.

Introduction

The increasing availability of detailed reaction mechanisms supports a growing interest in the engineering of combustion chemistry, especially regarding incineration, pollutants, materials, inhibitors, etc. In this work, published values of laminar burning velocity (LBV) of numerous gaseous fuel/oxidant mixtures are correlated with a calculated quantity based on the thermal diffusivity of the feed mixture and the space time at extinction (blowout) of a hypothetical perfectly stirred reactor (PSR). The blowouts are calculated using published detailed reaction mechanisms. Several cases are included which are model compounds for hazardous chlorinated chemical wastes.

Laminar Burning Velocity

A simple thermal treatment of the propagation of a laminar combustion wave through a gaseous fuel/oxidant mixture (Fig. 1) was considered by Mallard and Le Chatelier in the 19th century.

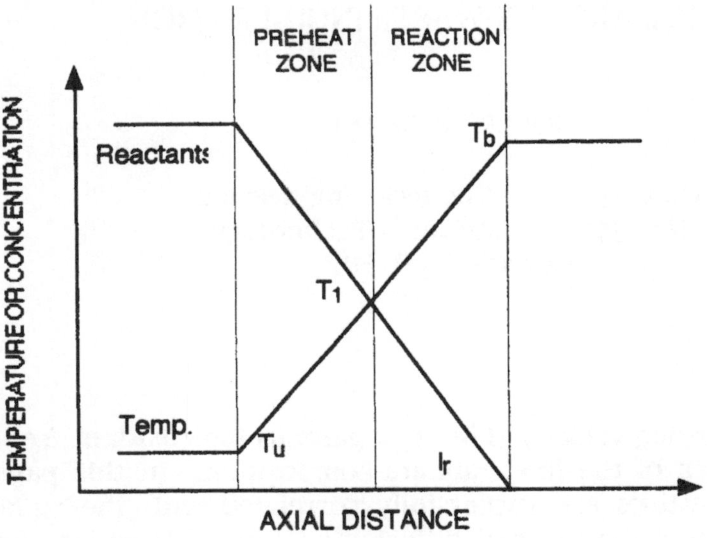

Figure 1: Schematic of plane laminar combustion wave

As described by Glassman (1987), the incoming fuel/oxidant mixture is heated in the preheat zone by conduction from the reaction zone. Using a linear approximation to the temperature gradient in the reaction zone, a simple energy balance yields:

$$k\frac{T_b - T_1}{l_r} = mC_p(T_1 - T_u) \tag{1}$$

where k = the gas thermal conductivity, T_b = the burned gas temperature, T_1 = an ignition temperature, m = the mass flow rate, T_u = the unburned gas temperature, C_p = a mean gas heat capacity, and l_r = the reaction zone thickness, which can be given by: $l_r = v_u t_r$, where t_r = the time in the reaction zone, and v_u = velocity of the unburned gases.

Since the flow cross section can be taken as constant, continuity provides that $m = \rho_u v_u$, where ρ = gas density. By definition, the laminar burning velocity $S_L = v_u$. Substitution into Eq. (1) yields:

$$S_L = \left[\frac{k}{\rho_u C_p}\frac{T_b - T_1}{T_1 - T_u}\frac{1}{t_r}\right]^{0.5} \tag{2a}$$

The group $(k/\rho C_p)$ is the thermal diffusivity a. The quantity $1/t_r$ is a characteristic reaction rate $\mathcal{R}$. The exponential temperature dependence of $\mathcal{R}$ dominates the linear temperature term. The simple thermal analysis, then, holds that the LBV is proportional to the square root of the product of the thermal diffusivity of the feed gas mixture and the combustion reaction rate:

$$S_L \propto [a\mathcal{R}]^{0.5} \tag{2b}$$

The LBV of a gaseous fuel/oxidant mixture is a good indicator of its relative combustion rate and stability. In order to utilize Eq. 2b to correlate published LBV values, a convenient method is needed to estimate the reaction rate $\mathcal{R}$ values. Stirred reactor theory will be used for this purpose.

Perfectly Stirred Reactor

In a perfectly stirred reactor (PSR), the feed is instantaneously, completely, and continuously mixed with the reacting fluid. The exit fluid composition and temperature are exactly those existing in the PSR. The utility of the PSR lies with its assumption of infinitely fast mixing relative to chemical rates. For a PSR, the species balances are:

$$\dot{m}(Y_k - Y_k^*) - \dot{\omega}_k W_k V = 0 \tag{3a}$$

where m = mass flow rate, Y_k = mass fraction of species k, * represents the feed, ω_k = net reaction rate of k , W_k = molecular weight of k, and V = PSR volume. The enthalpy balance is:

$$\dot{m}\sum_{1}^{K}\left(Y_k h_k - Y_k^* h_k^*\right) + \dot{Q} = 0 \qquad (3b)$$

where h_k = enthalpy of species k, Q = external heat loss rate, and K = total number of species. With a user-supplied reaction mechanism, this system of equations can be solved using the Chemkin PSR computer package (Glarborg et al., 1986). For a given mass flow rate, a stable PSR operating condition exists at the high temperature solution T, Y_k. It is well known (e.g. Barat, 2001) that, as the mass flow rate at constant feed composition is increased, an extinction ("blowout") condition is reached.

PSR and LBV

Stirred reactor theory has been used to model turbulent premixed flame stabilization in the wake of a bluff body, such as a rod normal to the cold gas flow. Glassman (1987) shows that the unburned gas velocity U_{bo} at the point of extinction (blowout) of a rod (diameter d) stabilized flame is related to the LBV of the fuel/oxidant mixture:

$$S_L \propto \left[a\frac{U_{bo}}{d}\right]^{0.5} \qquad (4)$$

Barat (2001) showed that experimental LBV values can be correlated according to a modified form of Eq. (4):

$$S_L \propto \left[\frac{k}{\rho_u C_p}\frac{w_{bo}}{\rho V}\right]^{0.5} = \left[\frac{a}{\tau_{bo}}\right]^{0.5} \qquad (5)$$

where w_{bo} = the calculated PSR blowout mass rate. The space time τ for the PSR at blowout, based on the high temperature exit condition, is given by $\rho V/w_{bo}$.

Numerous systems were studied (Barat, 2001), including nitrogen-containing species, hydrocarbons, chlorocarbons, and non-air oxidants. A total of 46 individual cases were run in order to cover a wide range of fuels, oxidants, feed conditions, and LBVs. The positive, linear correlation of the square root of the ratio of feed thermal diffusivity to hypothetical PSR space time at blowout with experimentally observed LBV (Eq. 5) accounts for the following:

- Wide range of LBV values (18 - 890 cm/s)
- Wide range of feed thermal diffusivities (0.038 - 2.2)
- Range of hydrocarbon fuels (C_1, C_2, C_3, C_6)
- Oxyhydrocarbon fuel (CH_3OH)
- Fuel mixture (Town Gas)
- N-based fuels (NH_3, HCN)
- Range of diluents (CO_2, N_2, He)
- Range of fuel equivalence ratios[ϕ] (lean-stoichiometric-rich)
- Inhibition by chlorocarbons (CH_3Cl, CH_2Cl_2)
- Non-O_2 oxidants (NO_2, N_2O)
- Range of pressures (0.1 - 5.8 atm)
- Range of fast H_2-based fuel systems
- Slow CO-based systems

he thermal diffusivity and mass density were calculated for the ed gas (fuel + oxidant) at 300 K and the system pressure. The mbustion of various fuel/oxidant pairs was treated with the PSR de and detailed reaction mechanisms taken from the literature. alculated PSR blowout space times based in exit conditions were etermined, and then correlated according to Eq. (5) with published 3V values. The adiabatic hypothetical PSR has a volume was 250 m^3, with a 300 K gaseous feed. The PSR calculations containing $_1$/C_2 hydrocarbon and nitrogen-based species were simulated with e Gas Research Institute (GRI) mechanism v. 2.11 (GRI, 1994).

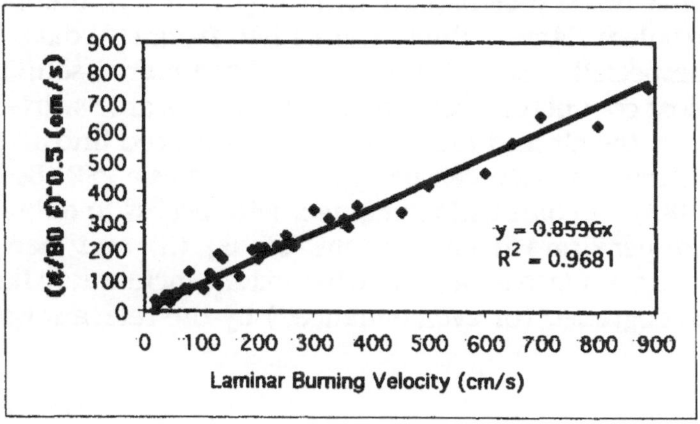

Figure 2: Correlation by Barat (2001) of LBV data with Eq. (5)

ϕ = (fuel/oxidant)$_{actual}$/(fuel/oxidant)$_{stoichiometric}$ based on molar or volumetric flow rates

For systems containing chlorinated species, the sub-mechanism of Ho et al. (1992) was attached to the GRI listing. For C_{3+} hydrocarbon systems, the mechanism of Castaldi et al. (1996) was used.

Utility for Hazardous Waste Incineration

The LBV varies with fuel equivalence ratio (ϕ), typically reaching a maximum at ϕ approximately one for hydrocarbon/air systems (Fristrom, 1995). Burning velocities are also quite dependent on levels of dopants, such as halogens (Westbrook, 1983). For example, CH_4/air flames doped with C_1 chlorocarbons show a marked decrease in maximum burning velocities as the chlorine content of the fuel increases (Gupta and Valeiras, 1984). This has significant implications for hazardous waste incineration since chlorinated solvents and sludges constitute a large mass fraction of the waste stream. Several researchers (Brouwer et al., 1992; Mao and Barat, 1996; Sacchi et al., 1996) have shown that increasing the chlorine content of a hydrocarbon/air incineration feed leads to increased emissions of products of incomplete combustion (PICs) and decreased combustion stability. This is directly related to a decrease in LBV as the chlorine content rises.

Commercial-scale combustors and incinerators operate at large, turbulent flow rates. However, turbulent flame speeds are directly related to laminar flame speeds (Beer and Chigier, 1983); therefore, insight into turbulent flame stability can be had from LBV data. This would be especially useful for mixtures. Even more useful is the fact that most commercial-scale turbulent flames are "swirl-stabilized". Here, the air and fuel/waste are introduced into a highly mixed flame zone that can be approximated as a PSR (Beer and Chigier, 1983). A combustion engineer with access to detailed mechanisms can perform PSR calculations, use Eq. (5), and then predict, at least on a relative basis, to what extent incinerator flame stability will be degraded (or even enhanced) by the relevant waste.

References

Beer, J. M. and Chigier, N. A. (1983). Combustion Aerodynamics. Malabar, FL: Krieger Publishing.

Brouwer, J., Longwell, J. P., Sarofim, A., Barat, R. B., & Bozzelli, J. W. (1992). Chlorocarbon-Induced Incomplete Combustion in a Jet-Stirred Reactor. Combustion Science and Technology, 85, 87-100.

Castaldi, M. J., Marinov, N. M., Melius, C. F., Huang, J., Senkan, S. M., Pitz, W. J., & Westbrook, C. K. (1996). Twenty Sixth Symposium (Int.) on Combustion (p. 693-702). Pittsburgh: Combustion Institute.

Fristrom, R. M. (1995). Flame Structure and Processes. New York: Oxford University Press.

Gas Research Institute – GRI (1994). http://www.gri.org.

Glarborg, P., Kee, R. J., Grcar, J. F., & Miller, J. A. (1986). PSR: A Fortran Program for Modeling Well-Stirred Reactors. Sandia Report SAND 86-8209. Livermore, CA: Sandia National Laboratories.

Glassman, I. (1987). Combustion, Second Edition. New York: Academic Press.

Gupta, A. K. & Valeiras, H. A. (1984). Burning Velocities of Chlorinated Hydrocarbon-Methane-Air Mixtures. Combustion and Flame, 55, 245-254.

Ho, W. P., Barat, R. B., & Bozzelli, J. W. (1992). Thermal Reactions of CH_2Cl_2 in H_2/O_2 Mixtures: Implications for Chlorine Inhibition of CO Conversion to CO_2. Combustion and Flame, 88, 265-295.

Mao, F. & Barat, R. B. (1996). The Interaction of Fuel-Bound Nitrogen and Fuel-Bound Chlorine During Air-Staged Combustion. Combustion Science and Technology, 116-117, pp. 339-357.

Sacchi, G., Procaccini, C., Longwell, J. P., Sarofim, A., (1996). Hazardous Waste and Hazardous Materials, vol. 13, no. 1, p. 39.

Westbrook, C. K. (1983). Numerical Modeling of Flame Inhibition by CF3Br. Combustion Science and Technology, 34, 201-225.

CATALYTIC COMBUSTION AS A POLLUTION PREVENTION TECHNOLOGY TO ACHIEVE ULTRA-LOW EMISSIONS IN POWER GENERATING GROUND-BASED GAS TURBINE ENGINES

Mitchell O. Stokes, Marco J. Castaldi*, Lance L. Smith, Hasan Karim, Shahrokh Etemad and William C. Pfefferle

Precision Combustion, Inc.
410 Sackett Point Road
North Haven, CT 06473

INTRODUCTION

In this paper it will be shown that there is significant motivation for – and advantages of – using catalytic combustion to improve gas turbine engine emissions and performance. In the discussion that follows, a brief account of the current state of gas turbine combustion will be given, followed by predicted calculations showing the need for catalytic combustion, and finally the presentation of actual data for a prototype catalytic reactor module operating at real gas turbine engine conditions, confirming that catalytic combustion has advantage over traditional lean premix combustion.

BACKGROUND

The Gas Turbine

In their most familiar application, gas turbines provide thrust for jet aircraft. In addition, they are also frequently used in power generation applications for the production of electricity, as shown in Figure 1. The gas turbine is comprised of three main components: a *compressor,* 'C', a *combustor* and a *turbine,* 'T' (see Figure 1).

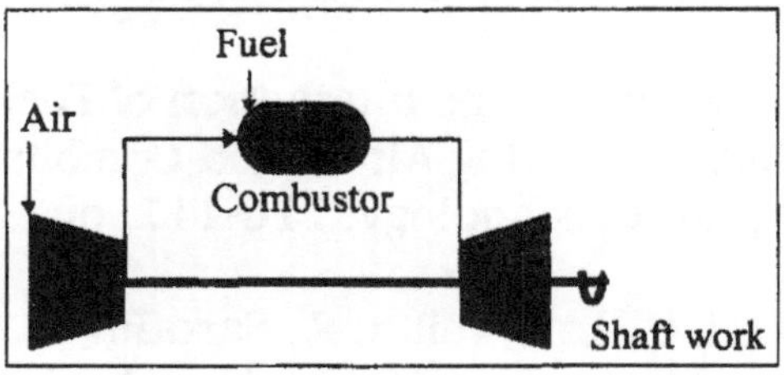

Figure 1. Schematic of a simple cycle gas turbine.

Air enters the compressor where it is compressed and consequently heated before it passes to the combustor. Within the combustor, fuel is injected into the air and combusted. Though the energy released by this reaction is used to turn

* Author for all correspondence – mcastaldi@precision-combustion.com

the turbine, not all of the energy is extracted making it necessary to either exhaust this higher-than-ambient temperature air into the atmosphere or further extract its energy in some subsequent process.

Conventional Gas Turbine Combustion: Diffusion Flames

Conventional gas turbine combustors typically operate in diffusion flame mode. In such a combustor, fuel is injected into the combustor without premixing with the combustion air. The injected fuel ignites and burns on contact with hot combustion gases in the combustor at very high flame temperatures (roughly 2260°C/4100°F for natural gas). And in addition to stabilizing the flame, this high combustion temperature allows most of the carbon monoxide (CO) and unburned hydrocarbons (UHC's) to burn out.

On the other hand, with high temperatures also come drawbacks. For example, the mechanism by which much of the resulting NOx in standard gas turbine combustors – in the form of nitric oxide or NO – is produced, is highly sensitive to temperature. This mechanism, sometimes called "thermal NOx" (or the "Zeldovich mechanism"), begins to rapidly produce NO at temperatures above 1525°C/2780°F.[1]

To minimize NOx, it is therefore necessary to keep the local temperatures beneath the 1525°C/2780°F levels but high enough to stabilize the flame and burn out CO and UHC's. This small window of temperatures forces gas turbine designers to tightly control the local fuel-air concentrations. This control has been sought by mixing the fuel and air prior to combustion – by premixing the fuel and air. There are, of course, other control methods for limiting harmful emissions – but most of them have been found economically wanting in terms of cost per unit of emission reduction. For example, water or steam injection requires costly auxiliary equipment. Exhaust after-treatment methods such as selective catalytic reduction (SCR) and selective non-catalytic reduction (SNCR) require large capital investments as well as high maintenance costs; in addition, ammonia slip can be an ever-present danger.

Lean Premix Prevaporized Combustion

In a diffusion flame, the fuel and air *locally* react as they come into contact with each other. So, even though the theoretical *overall* fuel-air mixture is much leaner than stoichiometric, the local fuel-air mixture at which the reaction actually occurs, is stoichiometric, and thus at correspondingly high temperatures. If the fuel and air are mixed prior to combustion, the resulting temperature will be lower than the stoichiometric temperature. In the limiting case in which the local fuel-air ratio is identical to the overall ratio (i.e. perfect mixing) the local temperature will correspond to the overall fuel-air ratio.

[1] The Zeldovich mechanism is

$$O_2 \Leftrightarrow 2O$$
$$O + N_2 \Leftrightarrow NO + N$$
$$N + O_2 \Leftrightarrow NO + O$$

Premixing is also relatively inexpensive compared to exhaust after treatment and water injection. There are drawbacks to premixing, however. The flame is generally less stable, CO and UHC's higher, and the premixer increases the pressure drop. In addition, there is the risk of auto-ignition and flashback.

Figure 2 below shows the effects of mixing on the production of NOx for methane and air at a range of adiabatic flame temperatures. These results were generated by incorporating the Chemkin II chemical kinetics package[1] along with the Gas Research Institute (GRI) chemical reaction mechanism for methane combustion[2]. The reaction was then modeled in two steps. The first step models a recirculation zone using a perfectly stirred reactor (PSR) model[3] while in the second step the flow structure is modeled as a plug flow reactor (PFR)[4]. For the plot shown, the PSR residence time was 0.5 ms and the PFR residence time 15ms for a *nominal* reaction time of "16" ms. The level of mixing was simulated by appropriately weighting the NOx over a discretized domain assuming a Gaussian distribution.

The "unmixedness parameter," *S*, is defined as the standard deviation of the local fuel-air ratios (over an assumed discretized domain) divided by the overall fuel ratio. To calculate the overall NOx for a given level of mixing, a Gaussian distribution of the local NOx values was assumed. Note that as the mixing improves, S gets smaller; with perfect mixing corresponding to an S of 0%. A diffusion flame, on the other hand, corresponds to an S=100% - that is, where no mixing takes place before the reaction. Figure 2 confirms the following: as the level of mixing increases (i.e. S gets smaller) the production of NOx decreases. More specifically, even as little as 25% unmixedness results in substantial NOx concentrations. Today's gas turbines operate in the range of 1500°C to 1600°C where typical premixing levels result in approximately 25 ppm NOx. Yet even at these relatively high NOx levels the combustors are usually operating on the edge of their respective stability limits.

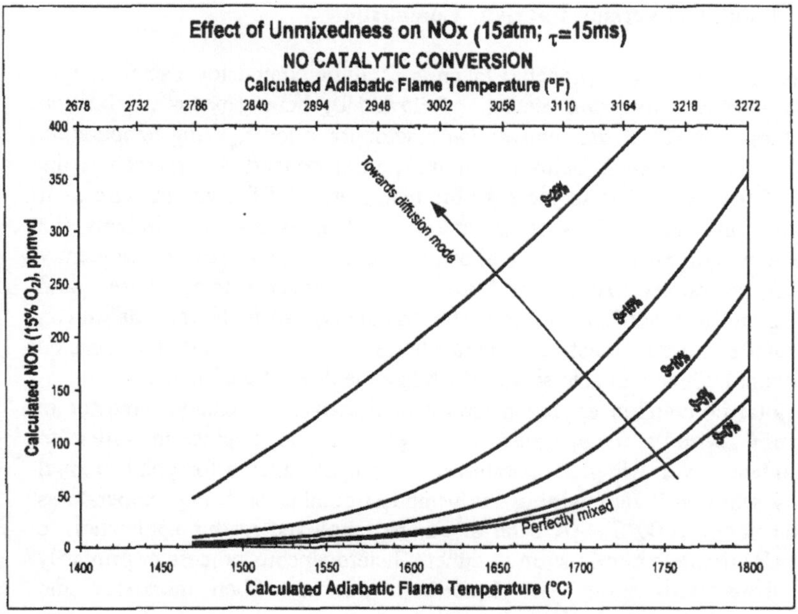

Figure 2. NOx vs. calculated adiabatic flame temperature showing the effect on unmixedness on NOx formation levels.

Catalytic Combustion

Given both the benefits and drawbacks of premix combustion, it is obviously desirable to find a method of combustion that gives similar NOx benefits while also providing flame stability. The following discussion will address the method of heterogeneous catalysis and its application to catalytic combustion as a way to strike the proper balance between NOx emissions on the one hand, and stability, CO, and UHC's on the other.

Now, while general heterogeneous catalysis technology is well understood, its application to catalytic combustion for gas turbines requires significant development effort. This further development is required because of the relatively high temperatures resulting from combustion in a gas turbine as well as the turbine's operational needs. For example, the catalytic process must contend with rapid heat release, high operating temperatures, variable fuel concentration and quality, variable inlet pressure and temperature, and variable reactor residence times (i.e. space velocity). But in exchange for such demands, the catalytic combustor is not generally required to maximize yield or selectivity to a particular product. It is required, however, to provide flame stability, usually by providing a temperature rise to a fuel/air mixture. Consequently, for catalytic combustion, the overriding constraint is *robustness* and *controlled operating temperature*.

Partial vs. Full Conversion Catalytic Combustion

There are two main approaches taken in catalytic combustor reactor design: (1) use a *partial conversion* reactor – followed by homogeneous combustion downstream – or (2) a *total conversion* reactor for a lower firing temperature combustor (see Figure 3 below). In the partial conversion application the purpose of the catalyst is to increase the temperature of the gas mixture as it enters the combustor. This temperature rise then assists in stabilizing the homogeneous combustion thereby allowing leaner operation and consequently lower NO_x emissions. And in addition to this increase in temperature – and depending on the conditions of the reactor and the type of fuel – the catalyst can also provide more reactive intermediates (such as partially reacted hydrocarbons) which can also serve to stabilize the downstream flame.

The total conversion approach uses a total-conversion catalytic reactor in combination with a recuperated cycle gas turbine engine in order to accommodate lower firing temperatures. This cycle allows for good thermal efficiency while still maintaining fairly modest combustor firing temperatures (less than about 1000°C). A catalyst can be employed in this application to accomplish complete combustion in either a heterogeneous process or primarily heterogeneous with some gas phase burn out of carbon monoxide and hydrocarbons. Thermal NOx formation is avoided because of the low combustor operating temperature. The disadvantage is that he catalyst must operate at the combustion temperatures.

Since partial catalytic combustion affords a much wider range of applicability, only this approach will be discussed here. Now, under the category of 'partial conversion' PCI has implemented several technologies to effect combustion stability and emissions. The emission and stability data presented below will show the development of one of these technologies.

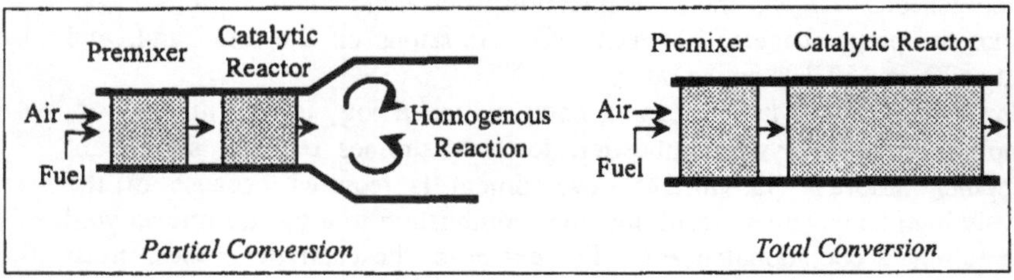

Figure 3. Schematics of partial (left) and total (right) conversion catalytic combustors.

RESULTS AND DISCUSSION

Test Conditions

A series of experiments were conducted in which different catalyst formulations, geometries, flowrates and conditions were tested to evaluate their effect on catalytic reactor performance. The conditions included fuel-air ratios from 0.017 to 0.041 and inlet temperatures from < 400°C to 600°C.

The catalyst surface temperature was monitored during the testing through K-type thermocouple measurements. For various inlet temperatures, the lean limit of operation was established by lowering the fuel flow to the point of notable increase in carbon monoxide emissions and/or flame extinction.

Catalyst Development

The catalyst should be active to "light-off" over a wide range of fuel-air stoichiometries. This will ensure that adequate heat release occurs through the catalytic reactor, to ensure acceptable downstream flame stability. Catalyst light-off temperature is a function of many variables, usually those associated with the operating parameters of the engine (e.g. velocity or residence time across the catalyst, fuel-air concentration, fuel quality/composition, as well as ambient air composition. Higher amounts of impurities in the fuel (such as sulfur) result in higher lightoff temperatures for the catalyst.

While high temperatures are potentially harmful to the catalyst and substrate, quick and extreme changes in temperature (e.g. at light off and shut down) can create the potential for severe thermal shock. Therefore, metallic substrates are preferable over ceramics since they readily accommodate thermal expansion. The use of metal substrates require that high temperature metals be used.

Catalyst testing involves the development of catalyst and substrate formulations for the reactor module. This includes new formulation and processes development in the chemistry laboratory addressing adhesion, materials and process compatibility as well as microstructural characterization of the catalyst and substrate. PCI practices catalyst screening on thin strips (or "coupons") and catalyst activity/performance testing in a reactor at both atmospheric and high pressure (10-15 atm) conditions.

For the catalyst screening, catalyst/substrate combinations are applied to flat, thin strips (or "coupons") for rapid testing. The coupons are then installed in a simple test rig, and exposed to a flow of premixed, preheated fuel and air. Catalyst performance is evaluated by the surface temperature rise due to catalytic surface reactions, as measured by a type K thermocouple. Temperature rise is recorded as a function of inlet fuel/air temperature and mixture ratio. Gas chromatograph (GC) analysis of the pre- and post-catalyst streams is also obtained, to provide a measure of fuel conversion and reaction rate.

Following coupon testing, promising catalysts are tested at simulated operating conditions, in a sub-scale reactor at both atmospheric pressure and high (10 atm) pressure. Results for five typical catalyst formulations are shown in Figure 4. Here, the performance of five catalyst formulations (labeled "A" through "E") are compared on the basis of catalyst surface temperature (labeled "Tsurf"), lightoff temperature (labeled "T l-o"), and gas temperature increase (labeled "delta T").

The strategy of formulation development and selection is initially focused on meeting the light -off temperature, T l-o. Tl-o has to be lower than the combustor inlet operating temperature by a specified margin to assure catalyst activity. Having satisfied this criteria, the reactor and formulation development should attain a design that operate the reactor surface temperature, Tsurf. Tsurf should be in a moderate temperature region defined by the material specification. This is important in terms of attaining long term reactor durability and meeting cyclic environment of start-up and shut down of the gas turbine engine. Finally, the reactor should provide sufficient gas temperature increase, delta T, to assure downstream flame stability and the resulting low emission operation.

Extensive testing was also performed at high pressure (10 atm) on three catalyst formulations. Temperature profiles axially through the reactor (stations 1 through 9) are shown for the three catalyst formulations in Figure 5. In general, surface temperatures were moderate and within the material limit temperature, with catalyst formulation "C" showing the best performance in terms of steady well-moderated operating temperatures throughout the full length of the reactor.

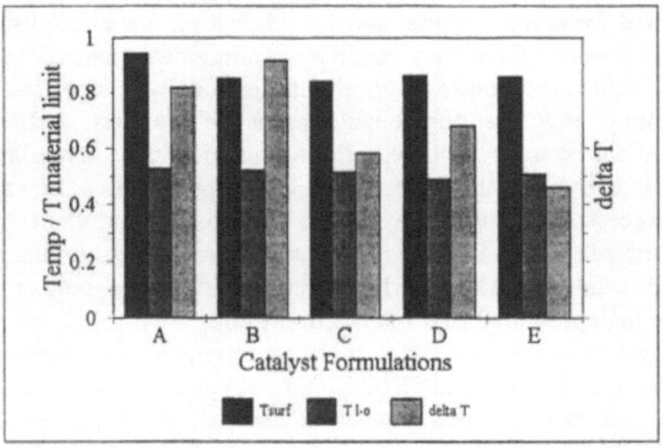

Figure 4. Comparison of catalyst performance data obtained for five catalyst formulations at 10 atmospheres (labeled A through E).

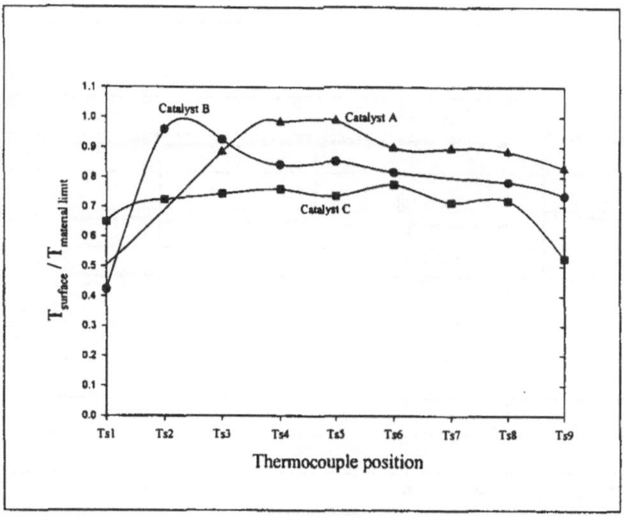

Figure 5. Catalyst temperature profiles in the reactor operating at 10 atm pressure, for A, B, and C catalyst formulations.

Catalyst Durability Testing

Limited durability testing (10 to 50 hours) was conducted to evaluate the initial stability of catalyst formulations. The most promising formulation has been tested for approximately 500 hours at 15-atm.

The 500-hour high-pressure data are shown in Figure 6. During this test, catalyst surface temperatures were maintained at a constant value, to allow interpretation and extrapolation of post-test materials analysis data. Performance testing and rig stabilization during the first 50 hours of testing led to some variation in catalyst surface temperature, as seen in the figure. Despite the slight variations in operating conditions, the data clearly show no degradation in catalyst performance during the 500-hour high-pressure test. It should also be noted that performance testing was conducted both prior and subsequent to the durability testing period shown – over a range of conditions – to bring the total run time on the catalyst to approximately 500 hours. The performance testing showed stable reactor performance.

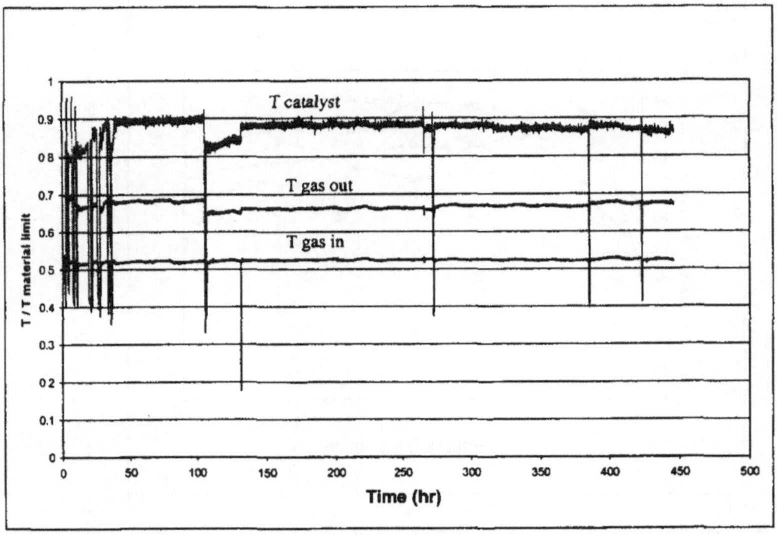

Figure 6. Catalyst surface temperature (and gas inlet and exit temperature) versus time, from 500-hour high-pressure (15 atm) durability test of the reactor.

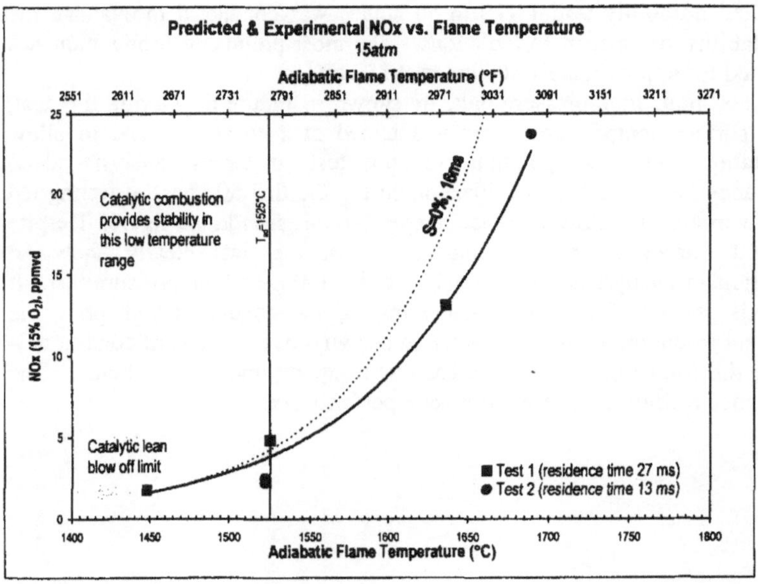

Figure 7. Plot of calculated and measured NOx emissions (corrected to 15% O_2) as a function of adiabatic flame temperature

High-Pressure Emission Testing

Exhaust emissions including UHC, CO, CO_2, O_2 and NO_x were measured at a number of different steady state conditions. A water-cooled rake type sampling probe was designed and built to provide six zones of equal area exhaust gas sampling. The rake was located 0.4 m (1.3 ft) downstream of the catalytic reactor exit, providing a gas residence time of 50-100 ms from the catalytic reactor exit to the rake type sampling probe for the range of velocities tested. A sample delivery pump, calibrated for the desired amount of sample gas flow, was used to draw the exhaust sample through the sampling probe and into the analytical instruments for emissions detection.

The emissions system used for the testing consists of a flame ionization detector based total unburned hydrocarbon (UHC) analyzer, nondispersive infrared detector for carbon monoxide (CO), a nondispersive infrared carbon dioxide (CO_2) detector, a paramagnetic oxygen (O_2) detector, and a chemiluminescent analyzer for oxides of nitrogen (NO_x). Each instrument was mounted in a fully portable rack, with pumps and back- pressure valves to maintain constant pressure and flow to all instruments. A thermoelectric chiller was used behind the UHC analyzer to cool (to 4°C) the heated sample gas and condense out the water before delivery to the other instruments. The instruments were zeroed and spanned with bottled calibration gases before each test series to assure accurate and precise measurements. Data were obtained directly from the instruments and recorded.

Fifteen atmosphere pressure data, including downstream gas-phase burnout, were obtained for the catalytic combustor module. The data confirm the low-NOx capability of the system, as well as operation at high firing temperature shown in Figure 7. Here, 15-atm data show the sub-scale NOx output (corrected to 15% O_2) as a function of final adiabatic flame temperature. For all data points, combustion is stabilized by catalytic pre-reaction from the reactor module. Large turndown from low, lean blowout to high firing temperature was achieved. The gas sampling port was located at a position that provided an estimated 13 ms residence time in the burnout zone, prior to sampling. PCI's chemical reaction modeling results for NOx formation are also shown, for the previously-discussed case (a PSR followed by a PFR corresponding to a nominal 16 ms). Low single digit NOx emissions are obtained at flame temperatures below approximately 2800°F (1530°C).

While the agreement between the predicted and the experimental values is good for the low adiabatic temperatures, there is discrepancy at the higher temperatures. This may be explained in two ways. First, an assumption of a .5 ms residence time was made for the PSR model. Second, it is known at the higher temperatures the calculated adiabatic flame temperatures are usually higher than those that are actually in the combustor due to greater amounts of heat loss from the system at high temperatures.

The catalytic reaction prior to gas-phase combustion provides flame stabilization at low adiabatic flame temperatures. To investigate this stability he lean blowout, or flame stability limit, was tested. This was done by slowly reducing the fuel until the CO concentration increased to very high levels and/or the flame extinguished. Test results for the catalytic module showed lean blowout occurred with 20% less fuel than the lean blowout point without catalytic pre-reaction.

CONCLUSION

The motivation for and advantages of using catalytic combustion to improve gas turbine engine emissions and performance has been shown for a prototype catalytic reactor module operating at real gas turbine engine conditions of 10 to 15 atmospheres. In addition, ultra-low NOx emissions with simultaneous stable combustion at up to 1525°C were achieved. Sub-scale reactors enabled extensive testing and experimentation to be done which yielded insights on the catalyst performance effect on combustion stability (extending lean blowout) and emission formation (ability to achieve low NO_x, CO and UHC). Current NO_x modeling has been validated by both sub- and full-scale engine prototype testing. Finally, a selected catalyst formulation has been tested for 500 hours under simulated engine conditions with no performance degradation. Current efforts are underway to develop an extrapolation model to predict the life of the catalyst when the reactor module is in field operation with the initial milestone of 8000 hours operation.[2]

REFERENCES

1. Kee, R.J.; Rupley, F.M.; Miller, J.A.; Sandia National Laboratories Report # SAND89-8009B; Sandia National Laboratories, 1995.
2. Frenklach, M.; et al.; Gas Research Institute Report #GRI-95/0058; Gas Research Institute, 1995.
3. Glarborg, P.; et al.; Sandia National Laboratories Report # SAND86-8209; Sandia National Laboratories, 1992.
4. Lutz, A.E.; Kee, R.J.; Miller, J.A.; Sandia National Laboratories Report # SAND87-8248; Sandia National Laboratories, 1996.

[2] PCI would like to acknowledge Siemens Westinghouse Power Corporation for high-pressure 15-atmosphere rig time. PCI thanks Mr. Donald Newburry and Dr. Tom Lippert from Siemens Westinghouse Power Corporation for providing valuable technical information and support throughout the project. Dr Richard Johnson's (program monitor from DOE-NETL) guidance throughout the project was also invaluable.

The Development of Gasoline Fuel Processors for Use in PEM Fuel Cell Vehicles

Antonio M. Vincitore
International Fuel Cells
South Windsor CT, 06074

As economic and environmental constraints generate a greater need for more efficient and cleaner power generation processes, Proton Exchange Membrane (PEM) fuel cells have become an area of aggressive research. However, the lack of a hydrogen fuel infrastructure poses a great problem for the near-term success of PEM fuel cells. As a result, numerous development efforts are being made to establish key fuel processing strategies that can satisfy on-demand hydrogen generation. Key focus is on developing a gasoline fuel processor that is capable of processing pump-grade gasoline into fuel cell grade hydrogen. Test results of an ambient-pressure gasoline fuel cell power plant will demonstrate IFC's success in converting California Reformulated Phase II gasoline to electricity.

SYMPOSIUM 10:

Industrial and Wastewater Treatment

A COMPARISON OF THREE DEGRADATIVE PRE-TREATMENT PROCESSES FOR DYE WASTES

MARK LYNCH
Centre for Applied Colloid and Biocolloid Science,
Swinburne University of Technology,
PO Box 218, Hawthorn, VIC 3122, Australia

VERONIQUÉ LÉVY
EWL Consultants, PO Box 39443,
Winnellie, NT, Australia

JOHN McCARTHY
Centre for Applied Colloid and Biocolloid Science,
PO Box 218, Hawthorn, VIC 3122, Australia

GREG LONERGAN
Centre for Applied Colloid and Biocolloid Science,
PO Box 218, Hawthorn, VIC 3122, Australia

ABSTRACT

Three options for oxidative pre-treatment of wastewater containing the triarlymethane dye Green S were studied and compared. Degradation of the dye was achieved using the fungal enzyme laccase, Fenton's reagent and ozonation. All three techniques were capable of reducing absorbance at λ_{max} by more than 98%. Biodegradability of the test solution was significantly enhanced by all three processes. The COD:CBOD$_5$ ratio was reduced from 275:1 to 11:1 by Fenton's reagent and to 2:1 by laccase and ozone. Treated solutions from all three processes showed no acute toxicity toward un-acclimatised activated sludge biomass. ADMI colour values were reduced by 81%, 97.8%, and 99.7% by laccase, Fenton's reagent and ozonation respectively.

INTRODUCTION

Conventional physical and biological processes used by industry for wastewater treatment have only a limited ability to degrade a broad range of organic dye compounds [1]. When dye removal is necessary, conventional processes must be augmented with further treatment to either physically separate or to decolourise the dye. Pre-treatment processes designed to degrade dyes include chemical oxidation, photo-catalytic oxidation and biological degradation [2].

Although a significant amount of research work has been reported on individual pre-treatment options, little data appears to be available directly comparing options. This paper compares three oxidative pre-treatment processes, the fungal enzyme laccase, Fenton's reagent and ozone. These processes, which have been independently investigated by the authors for a variety of applications, all have the ability to oxidise recalcitrant organics [3,4,5]. The comparison was based on the ability of the three processes to decolourise a model compound, the triarylmethane dye Green S and on the biodegradability of the oxidation reaction products.

Fungal enzymes

Over the past decade, the white rot fungi have been studied for their ability to degrade recalcitrant organic pollutants. The low specificity of ligninolytic enzymes produced by the fungi enables a wide range of compounds to be degraded, including polyaromatic hydrocarbons, chlorophenols and polychlorinated biphenyls [6].

The enzymatic pre-treatment investigated uses *Pycnoporus cinnabarinus* laccase to oxidise the phenolic group of the dye. This one electron oxidation forms a phenoxy radical which is subsequently oxidised by laccase exposing the ring to nucleophilic attack by water [5]. Dye decolourisation using fungal enzymes is the least developed of the three processes studied but may have advantages over chemical alternatives in terms of environmental acceptability. The cost and practicality of using enzymes for full-scale waste dye decolourisation applications is very difficult to predict based on existing evidence.

Fenton's Reagent

Fenton's reagent is the combination of hydrogen peroxide with ferrous iron. The reagent forms hydroxyl radicals by reduction of the hydrogen peroxide and oxidation of the iron to the ferric form. Hydroxyl radicals are among the most powerful oxidising agents and are less selective than alternatives such as chlorine or ozone. The optimum pH conditions for Fenton's reagent is between 3 and 3.5 [7,8,9]. This process is of interest because it entails a relatively low capital cost but is known to be highly effective at degrading a wide range of aromatic compounds.

In the presence of aromatic compounds, Fenton's reagent initiates a free radical chain reaction mechanism [10]. The general process is described by the following reaction:

$$H_2O_2 + Fe^{2+} \longrightarrow Fe^{3+} + OH^- + OH^\bullet$$
$$RH + OH^\bullet \longrightarrow R^\bullet + H_2O$$
$$R^\bullet + Fe^{3+} \longrightarrow R^+ + Fe^{2+}$$
$$R^+ + OH^- \longrightarrow R\text{-}OH$$

Ozonation

Ozone is one of the most powerful known oxidising agents and is used to degrade organic contaminants present in water supplies and in wastewater [11]. Ozonation for dye decolourisation has been demonstrated at laboratory scale for a broad range of dyes [12,13]. A number of different ozone-dye reactions have been reported. These include addition reactions during which ozonides are formed and ultimately carbonyl compounds and peroxides. Insertion reactions, electrophilic and nucleophilic attack and electron transfer have also been reported [12]. Ozonation can also decolourise dyes by the reaction of free radicals formed as a result of ozone decomposition. This decomposition proceeds by way of the formation of $OH^{\bullet}$, $HO_2^{\bullet}$ and $HO_3^{\bullet}$.

Despite the success of laboratory and pilot scale trials, the widespread application of ozone to decolourise wastewaters has not materialised. The perceived problem with the process is relatively high capital and operational costs [1,14].

METHODS

The model dye

Green S is a polyaromatic compound that is used as a food colorant in some countries (Australia and the UK are examples), for dyeing wool, silk and leather and as a biological stain. The dye is of interest both as a model compound and as a component of some food manufacturing effluents. The dye used for this study was a commercial preparation obtained from Warner Jenkinson Australia Pty Ltd. All decolourisation experiments were performed using a dye concentration of 100 mg L^{-1}, as supplied. The characteristics of the dye are listed in Table I and Figure 1 shows the absorbance spectra for a 10 mg L^{-1} Green S solution at pH 7.0.

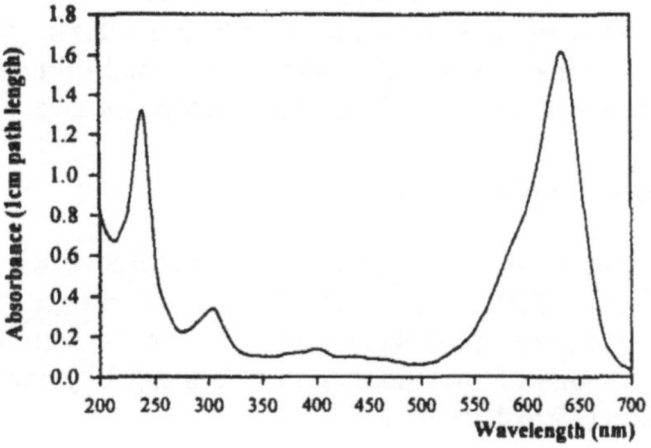

Figure 1. Absorbance spectra for Green S (10 mg l^{-1} as supplied at pH 7)

Table I. Key characteristics of Green S

<table>
<tr><td>Molecular formula</td><td>$C_{27}H_{25}N_2O_7S_2Na$</td></tr>
<tr><td>Structural formula</td><td>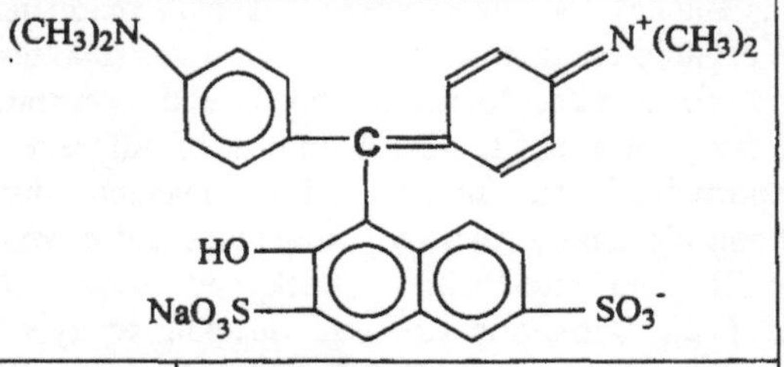
</td></tr>
<tr><td>Chromophore</td><td>Triarylmethane</td></tr>
<tr><td>Maximum absorbance wavelength
λ_{max} (nm)</td><td>632 at pH 7.0
615 at pH 9.0
640 at pH 4.5</td></tr>
<tr><td>Colour index number (C.I.#)</td><td>44090</td></tr>
<tr><td>Molecular weight (g mol^{-1})</td><td>576.6</td></tr>
</table>

Decolourisation using laccase

Laccase was produced using a 5.5 L continuously fed reactor inoculated with *Pycnoporus cinnabarinus* and operated at 25°C with no recycle. The feed solution [5] was filtered using 0.2μm PTFE membranes and fed at 0.5 L d^{-1}. Air, filtered using 0.2μm PTFE membranes, was used to saturate the reactor with dissolved oxygen. Extracellular fluid from the bioreactor was collected in sealed containers and the laccase activity was measured daily.

The decolourisation reaction was carried out by mixing the dye with concentrated *Pycnoporus cinnabarinus* extracellular fluid (laccase dose 1.0 U mL^{-1}) in a stirred batch reactor. A mediator (1-hydroxybenzotriazole) was added at 2 mmol L^{-1} and the reaction mixture was buffered using sodium acetate buffer (25 mmol L^{-1}, pH 4.5). The reaction mixture was saturated with oxygen by bubbling air through the mixture for the duration of the decolourisation trial and incubated at 45°C. These reaction conditions have previously been demonstrated as the optimum conditions for dye decolourisation using laccase [5]. Samples were taken every 15 minutes over a 120 minute period and diluted 1:10 with distilled water. Absorbance of the diluted samples was then measured at 632 nm.

Decolourisation using Fenton's reagent

The Fenton's reagent was produced using a freshly prepared stock solution of ferrous iron ($FeSO_4.7H_2O$, 2780 mg L^{-1}) and 30% (m/v) hydrogen peroxide. A ferrous iron and hydrogen peroxide dose rate was selected from an initial series of experiments using a range of iron doses (1.4 to 5.6 mg L^{-1}) and a range of hydrogen peroxide doses (39 to 156 mg L^{-1}).

The decolourisation reaction was carried out in a stirred batch reactor operating at room temperature. The dye solution was first adjusted to pH 3.5 using concentrated H_2SO_4 and NaOH. Ferrous iron was added (2.8 mg L^{-1}) and the reaction initiated by the addition of hydrogen peroxide (78 mg L^{-1}). Samples were taken every 15 minutes over a 120 minute period, diluted 1:10 with distilled water and the pH was raised to 9.0 to stop the oxidation reaction. Absorbance of the samples was then measured at 615nm.

Decolourisation using ozone

Ozone was produced by a UV ozone generator supplied by Australian Ultra Violet Services (AVUS). Air/ozone flow rate from the generator was 960 mL min^{-1} (at 20°C, 8kPa gauge pressure) and the output of O_3 was 90 mg h^{-1}. A 675mm tall, 65mm diameter contactor was used for the experiments. Residual gas from the contactor was passed through two gas washers connected in series that contained a 2% (w/w) KI solution and a phosphate buffer (pH 7). Titrametric measurement of iodine liberated in the gas washers was used to calculate the quantity of ozone that was not consumed in the contactor and so it was possible to calculate the ozone consumption.

The decolourisation reaction was studied using the Green S solution in a phosphate buffer (2.87 g L^{-1} KH_2PO_4, 6.59 g L^{-1} $K_2HPO_4.3H_2O$) to maintain a neutral pH. Prior to each decolourisation experiment, the ozone generator was operated for 60 minutes to allow ozone output to stabilise. The ozone / air flow was then diverted to the test contactor. At 30 minute intervals 10 mL samples of the solutions were drawn directly out of the contactor and corresponding samples of 100 mL were drawn out of the gas washers. The absorbance of the reaction mixture was determined by measurement of a diluted sample (1:10 into distilled water) at 632 nm.

Analytical procedures

Laccase activity was determined by measuring the oxidation of 300 μmol L^{-1} ABTS (2,2-azinobis-(3-ethylbenzthiazoline-6-sulphonate) buffered with 50 mmol L^{-1} sodium citrate (pH 3.5). Formation of the cation radical was monitored at 415 nm. Enzyme activities are expressed in Units (1 Unit, U, is the amount of enzyme required to oxidise 1 μmol of ABTS per minute).

Samples of pre-treated dye were analysed for $CBOD_5$ and COD [15]. Ozone generator output and ozone consumed in the gas washers was calculated using iodometry [15].

All absorbance measurements were carried out using a Varian Cary3 UV/Vis spectrophotometer. Absorbance at λ_{max} was used as a measure of the extent of decolourisation. Absorbance data from 400 – 700 nm were used to calculate the ADMI colour value [16,17]. The ADMI value is a measure of sample colour that is independent of hue. The colour value can be used to indicate the presence of coloured oxidation products with absorption properties that are different from the original dye.

The effect of the pre-treated dye solutions on the dissolved oxygen utilisation rate (DOUR) of activated sludge biomass was measured as an indication of acute toxicity. 100 mL of biomass (equivalent to 100 mg VSS) from an un-acclimatised laboratory scale activated sludge reactor was added to 50 mL of pre-treated dye solution. A feed solution was added (0.3 mg $CBOD_5$ per mg biomass), the mixture was diluted to 500 mL and the DOUR was measured [15].

RESULTS AND DISCUSSION

The optimal conditions for laccase pre-treatment of a range of commercial dyes have been determined to be: pH 4.5, temperature 45°C with a laccase mediator (1-HBT) concentration of 2 mmol L^{-1} [5]. With these optimised conditions the decolourisation of Green S proceeds rapidly (Figure 2). Absorbance at λ_{max} was reduced by 98.5% after 120 minutes. Maximum removal rate, expressed as percentage reduction in absorbance at λ_{max}, was 2.0% min^{-1}. ADMI colour value was reduced from to 17043 to 3214 (initial solution ADMI was 17854 at pH 7.6). Note that the rate of Green S decolourisation by laccase is relatively temperature independent between 25 and 45 °C [5] and although 45°C was used for these experiments, decolourisation is practicable at lower temperatures. This is obviously vitally important when considering the feasibility of full-scale applications.

The dose rates of ferrous iron and hydrogen peroxide used for this study were taken from a kinetic study of Green S decolourisation using Fenton's reagent [4]. A reduction in absorbance at λ_{max} that exceeded 99% was achieved with a hydrogen peroxide dose of 78 mg L^{-1} and a ferrous iron dose of 2.8 mg L^{-1} with a 120 minute reaction time. The maximum rate of reduction in absorbance at λ_{max} was 2.6 % min^{-1} (Figure 2). The ADMI value after 120 minutes was 363.

reduction in absorbance at λ_{max} greater than 99.9% was achieved after 90 minutes ozonation at a dose rate of 49.3 mg L h^{-1} (Figure 2), equivalent to an ozone consumption of 68.1 mg L^{-1}. After 90 minutes of ozonation, the dye solution retained a pale yellow colour and the ADMI colour value of the solution was 336. Following ozonation for 120 minutes, equivalent to an ozone consumption of 81 mg L^{-1}, the ADMI colour value was 172. An ozone consumption of 81 mg L^{-1} is equivalent to approximately 10 moles O_3 per mole of dye. The maximum rate of colour removal was 2.39% min^{-1}.

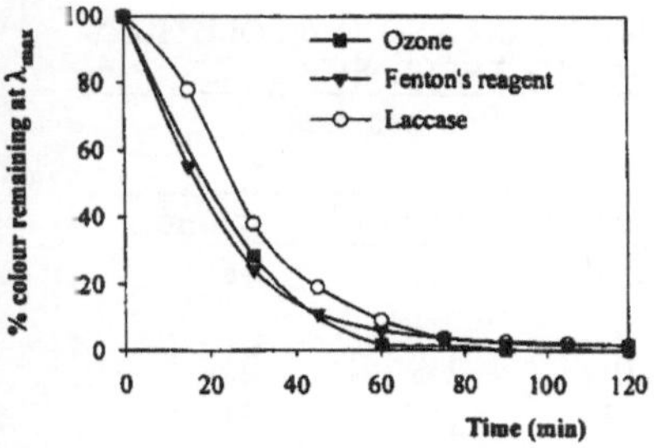

Figure 2. Percentage reduction in absorbance at λ_{max} by the three oxidants

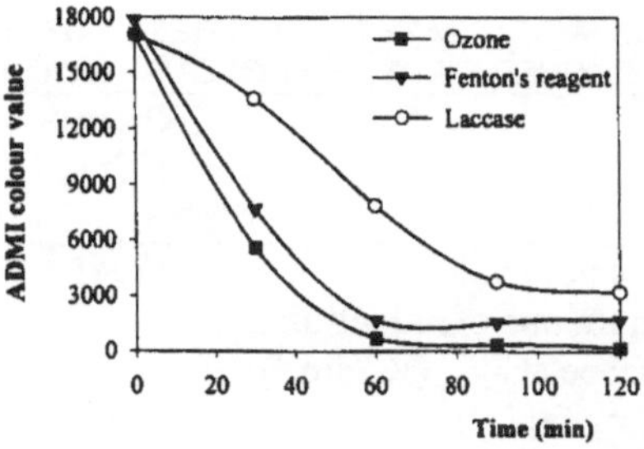

Figure 3. Variation of ADMI colour value by the three oxidants during the decolourisation reaction

The ADMI colour values for the treated solutions during decolourisation are compared in Figure 3. The final solution following laccase treatment had a residual colour value in excess of 3200. The crude enzyme used for these experiments contributed to the residual colour of the treated solution (crude enzyme colour value was 933). Residual colour in the solutions treated with Fenton's reagent may be caused by the formation of coloured iron complexes [10,18]. Whether or not modifications to the laccase and Fenton's pre-treatments can remove the residual colour is yet to be investigated. Only the ozonation process resulted in an ADMI colour value of less than 300 after 2 h (172 at 2h). This value is significant because some water utilities in Australia specify a colour standard for industrial discharge to sewer that is equivalent to an ADMI of 300 (after turbidity is removed).

Table II presents analytical data for the final solutions from the oxidative pre-treatments of Green S. In the case of ozone, the COD of the solution was reduced, however, the COD of the final solutions from laccase and Fenton's treatments increased. In the case of laccase, the increase in COD was attributed to the sodium acetate buffer and mediator added to the reaction solution. Control $CBOD_5$ tests using the buffer, enzyme and mediator show that this mixture contributes 109 mg L^{-1} $CBOD_5$. The residual $CBOD_5$ contributed by the dye degradation products is therefore 5mg L^{-1}. The increase in COD following Fenton's treatment may be caused by residual hydrogen peroxide in the sample, which is known to interfere with the standard COD test [10]. COD analysis of samples following Fenton's treatment conducted 24 and 48 h after decolourisation showed a progressively lower COD. This suggests that oxidation of the dye degradation products continues to occur, albeit at a very significantly reduced rate.

TABLE II. AGGREGATE PROPERTIES OF GREEN S SOLUTION AND THE SOLUTIONS PRODUCED BY DYE DECOLOURISATION USING THE THREE PROCESSES.

	Green S 100mg L^{-1}	Pre-treatment		
		Laccase	Fenton's	ozone
ADMI colour value	17043	3214	363	48
Maximum rate of reduction in absorbance at λ_{max} (% min^{-1})	-	2.0	2.6	2.4
COD (mg O_2 L^{-1})	110	237	155	41
$CBOD_5$ (mg L^{-1})	0.4	114	14.1	19.1
COD:$CBOD_5$ ratio	275:1	2.1:1	11:1	2.1:1
Effect on DOUR (% change)	+31%	+29%	+12%	+23%

In all cases, the COD to $CBOD_5$ ratio decreases very significantly, indicating that the reaction end products are more biodegradable than the dye solutions. Degradation processes operated simply to alter recalcitrant compounds, making them amenable to conventional biological treatment, are more cost effective than attempting complete mineralisation [19]. The combination of oxidative pre-treatment processes and biological post treatment has been suggested by a number of researchers as the optimum process combination for wastewater containing both recalcitrant and biodegradable organics [2]. Previous ozonation trials with a concentrated (500 mg L^{-1}) dye solution, have shown that the $CBOD_5$ increases to a maximum following an ozone dose of 68 mg L^{-1}, equivalent to 1.64 moles O_3 per mole dye. However, at this point in the reaction the solution was still very highly coloured (ADMI colour value was 20450 at pH 7.6). Further experimentation is required to determine the minimum oxidant dose required to adequately decolourise the dye solution using a two-stage process.

All degradation processes that do not lead to total mineralisation of organic components of the waste water may form reaction end products that are more toxic than the original components [20]. Consequently, it was important to determine whether the pre-treated solutions were likely to be inhibitory to subsequent biological treatment processes. The dissolved oxygen utilisation rate (DOUR) of biomass from a laboratory activated sludge system contained in a 500 mL incubation vessel was measured after addition of 50 mL of treated dye solution. The DOUR results are reported in table II as a percentage of the control values. In each case the DOUR increased. This suggests that the reaction products from each pre-treatment process are not acutely toxic to activated sludge biomass.

CONCLUSIONS

All three processes were capable of substantially decolourising the model dye. The COD:$CBOD_5$ ratio of the test solutions decreased significantly, indicating an increase in biodegradability. None of the pre-treated solutions were acutely toxic to activated sludge biomass (at the conditions tested). Residual colour resulted in ADMI values greater than 300 following pre-treatment by Fenton's reagent and laccase. It is possible that residual colour following Fenton's pre-treatment can be reduced by increasing the reaction time or oxidant dose rate. Further tests are needed to determine whether the residual colour in the laccase treated samples can be reduced by increasing the reaction time.

REFERENCES

1. Michelsen, D.L., W.W. Powell, R.M. Woodby, L.L. Fulk and G.D. Boardman. 1992. "Pretreatment of textile dye concentrates using Fenton's reagent and ozonation prior to biodegradation," presented at Chemical Oxidation Technologies for the Nineties: 2nd International Conference, Vanderbilt University, 1992.

2. Marco, A., S. Esplugas and G. Saum. *1997*. "How and why combine chemical and biological processes for wastewater treatment," *Wat. Sci. Tech., 35(4):321-327.*

3. Lynch, M.R., V.C.L. Lévy and G.T. Lonergan. 1998. "Degradation of dyes present in industrial wastewater using a combined chemical oxidation and biological treatment process," presented at Water Quality and Waste Treatment Conference, Lombok, Indonesia, October 1998.

4. Lévy, V.C.L., M. Lynch, J. McCarthy and G.T. Lonergan. 1999. "Oxidative pretreatment of waste food dye by Fenton's reagent," presented at Cleaner Production in the Food and Beverage Industries Conference, Hobart, Australia, September 1999.

5. McCarthy, J., V.C.L. Lévy, G. Lonergan and J. Fecondo. *2001*. "Development of optimal conditions for a range of industrial dyes using *Pycnoporus cinnabarinus* laccase," *Envir. Eng. Sc., (submitted).*

6. Reddy, A. *1995*. "The potential of white rot fungi in the treatment of pollutants," *Curr. Opin. Biotechnol. 6:320-328.*

7. Barbeni, M., C. Minero and E. Pelizzetti. *1987*. "Chemical degradation of chlorophenols with Fenton's reagent," *Chemosphere, 16:2225-2237.*

8. Debellefontaine, H. *1990*. "Le péroxyde d'hydrogène, agent d'oxydation des composés organiques en phase aqueuse," in: *Fondements Théoriques du Traitement Chimique Des Eaux*, H. Roques ed. Paris: Tech. Doc. Lavoisier, pp. 746-805.

9. Falcon M., B. Peyrille, P. Reilhac, J.N. Foussard and H. Debellefontaine. *1993*. "Oxydation en voie humide de la pollution organique aqueuse par le péroxyde d'hydrogène," *Rev. Sci. Eau., 6:411-426.*

10. Lin, S.H. and C.C. Lo. *1997*. "Fenton process for treatment of desizing wastewater," *Wat. Res., 31:2050-2056.*

11. White, G.C. *1992. Handbook of chlorination and alternative disinfectants 3rd edn.*. New York: Van Nostrand Reinhold.

12. Reife, A. and H.S. Freeman. *1996. Environmental Chemistry of Dyes and Pigments*. Washington DC: Wiley-Interscience.

13. Liakou, S., S. Pavlou and G. Lyberatos. *1997*. "Ozonation of azo dyes", *Wat. Sci. Tech., 35(4):279-286.*

14. Kallas, J. and R. Munter. *1994*. "Post-treatment of pulp and paper wastewaters using oxidation and adsorption processes", *Wat. Sci. Tech., 29(5-6):259–272.*

15. APHA / AWWA. *1995. Standard Methods for the Examination of Water and Wastewater 19th edn.*. Washington, DC: American Public Health Association/American Water Works Association/Water Environment Federation.

16. McLaren, K. *1970*. "The Adams-Nickerson color difference formula", *J. Soc. Dyers Colorists, 86:354-366.*

17. Allen, W., W.B. Prescot, J.R. Derby, C.E. Garland, J.M. Peret and M. Saltzman. *1973*. "Determination of color of water and wastewater by means of ADMI color values", *Proc. 28th Ind. Waste Conf., Purdue Univ., Eng. Ext. Ser. 142:661-675.*

18. Bandara, J., V. Nadtochenko, J. Kiwi and C. Pulgarin. *1997*. "Dynamic of oxidant addition as a parameter in the modelling of dye mineralization (Orange II) via advanced technologies", *Wat. Sci. Tech., 35(4):87-93.*

19. Scott, J.P. and D.F. Ollis. *1995*. "Integration of chemical and biological oxidation processes for water treatment: review and recommendations", *Env. Prog., 14(2):88–103.*

20. Gulyas, H. *1997*. "Processes for the removal of recalcitrant organics from industrial wastewaters", *Wat. Sci. Tech., 36(2-3):9-16.*

AUTOTROPHIC DENITRIFICATION WITH HYDROGEN OXIDIZING BACTERIA

KAUSER JAHAN
Associate Professor of Civil and Environmental Engineering
Rowan University
201 Mullica Hill Road
Glassboro, NJ 08028

TARIQ AHMED
TAMS Consultants
300 Broad Acres Drive
Bloomfield, NJ 07003

CLAIRE STEAGER
Graduate Research Assistant
Rowan University
201 Mullica Hill Road
Glassboro, NJ 08028

INTRODUCTION

Land disposal of secondary treated wastewater, septic tank effluents and agricultural runoff containing high concentrations of nitrogen compounds have led to an accumulation of nitrates in groundwaters in many areas of the USA [1-2]. Nitrate contamination of water sources is a serious problem due to the adverse health and ecological effects. Nitrates have been linked to infant methemoglobinemia (commonly known as the blue baby syndrome). N-nitroso compounds are also formed and known to be carcinogenic. Because of these reasons, the USEPA has set a Maximum Contamination Level (MCL) of 10 mg/L for nitrates in drinking water [3]. There are a number of physico-chemical and biological methods available for removing nitrate from water and wastewater. For drinking water, ion exchange is currently the predominant method for removing nitrate in the United States. Even with the recent development of nitrate selective resins, the maintenance and operating costs are still high, and the production of concentrated waste brines creates additional disposal problems. In European countries biological denitrification of drinking water using fixed bed reactors with ethanol or acetic acid as the carbon source is becoming more common [4-5].

Heterotrophic denitrification is a proven technology, but the residual organics can react with chlorine and form carcinogenic disinfection byproducts (DBPs). This has limited heterotrophic denitrification to wastewater applications in the United States because chlorine is the predominant means for disinfection of potable water. In Europe, biological denitrification of drinking water is being accomplished by using ozone or UV disinfection coupled with expensive post-treatment to remove the added organic contaminant and biomass. At this point, a method that can remove nitrate without producing concentrated waste brines or require an organic substrate would be very attractive to the water and wastewater treatment industry.

Biological denitrification is the most common method of removing nitrogen from municipal and industrial wastewaters [6]. The denitrification reactors follow nitrification processes that convert organic nitrogen and ammonia to nitrate. The denitrification reactors can be either suspended growth or attached growth with the latter being favored because it minimizes the space required. The disadvantages of heterotrophic denitrification are the cost incurred for adding an external organic carbon source for denitrification and the post-treatment needed to remove the residual organic contaminant and biomass before final discharge.

To eliminate chemical contamination from adding organic carbon, autotrophic bacteria can be used for denitrification. The growth rate of autotrophic bacteria are typically slower than heterotrophic bacteria, therefore, lower concentrations of biomass and soluble microbial products can be expected in the reactor effluent. The post-treatment at a pilot scale plant in Germany using hydrogen-oxidizing autotrophic bacteria consisted of a counter current aeration column where oxygen is introduced to remove the remaining nitrogen and hydrogen, and to remove biomass, flocculation and filtration of the effluent is carried out with a two layer filter [7]. In contrast to heterotrophic denitrification, the only substrate added was hydrogen which is completely harmless to potable water and easily removed.

For autotrophic denitrification no organic carbon source is required, rather carbon dioxide or bicarbonate are used for cell synthesis. For the energy source autotrophic denitrification requires hydrogen or various reduced-sulfur compounds. The following stoichiometric relationships have been formulated for autotrophic denitrification with hydrogen, thiosulfate, and sulfide serving as the electron donors:

$$2NO_3^- + 5H_2 \rightarrow N_2 + 4H_20 + 2OH^- \quad [8]$$

$$5S_2O_3^= + 8NO_3^- + H_2O \rightarrow 4N_2 + 10SO_4^= + 2H^+ \quad [9]$$

$$5S^= + 8NO_3^- + 8H^+ \rightarrow 5SO4^= + 4N_2 + 4H_2O \quad [10]$$

Autotrophic organisms, e.g., *Microccus denitrificans*, will reduce nitrate to nitrite and nitrogen gas while oxidizing hydrogen gas to water. The reaction proceeds as follows [8]:

$$2NO_3^- + 2H_2 \rightarrow 2NO_2^- + 2H_2O$$

$$2NO_2^- + 3H_2 \rightarrow N_2 + 2H_2O + 2OH^-$$

Combining reactions 1 and 2 the overall reaction can be written as:

$$2NO_3^- + 5H_2 \rightarrow N_2 + 4H_20 + 2OH^-$$

From the above equations, 0.35 mg H_2/mg NO_3–N is theoretically required for complete reduction. The experimental values are :0.38 mg H_2/mg NO_3–N [8], in a bench-scale reactor and 0.40 mg H_2/mg NO_3–N [11] at a full-scale plant in Germany. From the overall reaction given above, 2 moles of OH^- are produced for every 2 moles of NO_3^- reduced, so alkaline conditions and NO_2^- accumulation at high pH can be expected. To prevent pH shifts and nitrite accumulation, CO_2 can be added to buffer the system as well as serve the inorganic carbon source for cell synthesis.

A study conducted in Switzerland evaluated in-situ denitrification using hydrogen [12]. Nitrate contaminated groundwater was withdrawn, amended with hydrogen, carbon dioxide, and phosphate, and reinjected into the aquifer. The nitrate concentration in the aquifer was reduced from 92 to 11 mg/L within four weeks. The fact that H_2 gas is inherently clean makes it the ideal reactant to denitrify drinking water. Besides not being an organic compound, hydrogen generates less excess biomass than the use of organic compounds, and if generated on-site is less expensive than the use of methanol, ethanol, and acetate based on bulk chemical cost [13].

This study focuses on the use of hydrogen oxidizing bacteria for autotrophic denitrification. Batch biodegradation studies were conducted to evaluate the kinetic parameters of the autotrophic process. The performance of the process was evaluated by measuring the influent and effluent nitrate concentrations, the biomass concentrations and operating pH.

MATERIALS AND METHODS

Acclimation reactors were constructed for developing a mixed culture for the batch studies. The mixed culture was obtained from an anoxic section of a local wastewater treatment plant (Winslow Township Wastewater Treatment Facility, Winslow, NJ). Acclimation experiments were carried out at room temperature (20±2°C). Deionized distilled water was used in all experiments.

Nitrate concentration was periodically monitored in the acclimation reactors using the Standard Method 4500-B [14], on a HACH DR 4000U (HACH Co., Loveland, CO) spectrophotometer. When NO_3 concentration depleted below 30% of the initial concentration, 25 mL of the reactor content was used to seed a subsequent reactor. Due to the presence of free hydrogen gas, the entire experimental setup was operated under a laboratory exhaust hood. Cell mass was measured as MLVSS using Standard Method [14].

A schematic of the batch experimental setup is presented in Figure 1. Reactors were also equipped for carbon dioxide gas delivery for experiments that used carbon-dioxide only as the sole carbon source. These experiments are stil in progress and will be reported elsewhere. All Plexiglas batch reactors were filled with 380-mL HACH BOD based buffer solution containing a nitrate ($NaNO_3$) concentration of approximately 100-mg/L. The reactors were seeded with 20-mL of mixed liquor from an acclimation reactor containing bacteria newly acclimated to carbonate (Na_2CO_3) as their carbon source. The four batch reactors were placed on a HACH BOD Trak (Loveland, CO) apparatus to facilitate mixing. They were purged with pure hydrogen gas every 4 hours with the aid of Burkert (Burkert, Orange CA) solenoid valves, which are controlled by programmable ChronTrol® (Model XT, ChronTrol Corp., San Diego, CA) timers. A control reactor was also maintained. This reactor did not receive any microorganisms. Reactors were kept well mixed and closed to the atmosphere to ensure anoxic conditions.

pH in the reactors was controlled by the addition of acid or alkali. pH control is important as the reactions generate alkali in the system [8] which leads to nitrite accumulation.

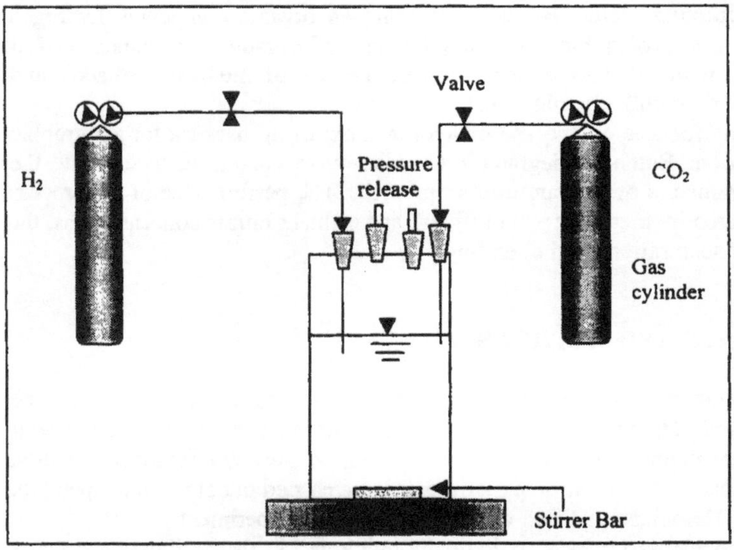

Figure 1: Schematic of Batch Reactors

Estimation of Denitrification Kinetic Constants

The Monod model [15] was used to describe the rate of nitrate removal and cell growth. The kinetic coefficients were obtained by fitting the simple Monod equation for substrate utilization and cell mass production. The rate of change of substrate and cell mass with time is given by:

$$\frac{dS}{dt} = -\frac{1}{Y}\frac{\mu_m SX}{K_s + S}$$

$$\frac{dX}{dt} = Y\frac{dS}{dt} - k_d * X$$

where,

$\frac{dS}{dt}$ = substrate conversion rate, mg/L-hr

$\frac{dX}{dt}$ = net rate of bacterial growth, mg VSS/L-hr

S = substrate concentration, mg/L

X = cell concentration, mg/L

K_s = half velocity constant, mg/L

Y = yield coefficient, mg VSS formed/mg substrate converted

k_d = decay coefficient, hr^{-1}

μ_m = maximum specific growth rate, hr^{-1}

For simplicity, this model assumes one reduction step from nitrate to nitrogen gas. Nitrate is modeled as the substrate (S). It was assumed that nitrite concentrations would be low because of adequate pH control. The basic assumptions of the model are:

a) microorganisms are well acclimated to the substrate
b) nutrients for balanced growth are not limited
c) there is no inhibition in the system
d) the growth is substrate limited

The above kinetic equations were solved simultaneously numerically for S and X using a sixth-order Runge-Kutta method. A search for the values of the kinetic parameters was carried out using an algorithm for least squares estimation. This algorithm finds a local minimum of the sum of the square weighted errors (SSWE).

RESULTS AND DISCUSSION

Nitrate removal was observed in all three reactors as presented in Figure 2. removal occurred in 50 hours for all reactors. Nitrate removal was not significant after fifty hours for all reactors. There was no nitrate concentration reductions in the control reactor during the duration of the experiment.

Cell mass could not be measured accurately as it was difficult to obtain representative samples from the batch reactors. The acclimated culture in the reactors showed a tendency to adhere to the reactor sides. Therefore cell mass as MLVSS was measured at the start and end of the experiment.

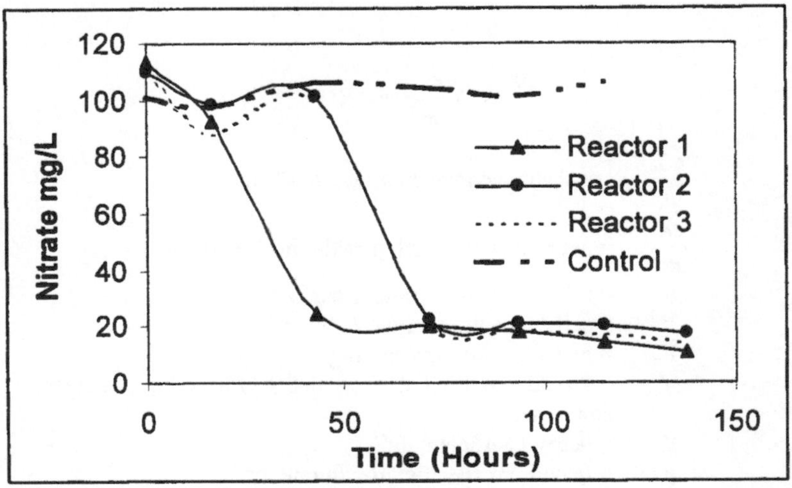

Figure 2: Nitrate Removal in Batch Reactors

The average of the nitrate removal in the three batch reactors was calculated and modeled using Monod kinetics [15]. The model was used to determone the best fit line for the experimental data. The results of the model prediction is shown in Figure 3. Cell mass data could not be obtained with time. However the initial and final cell mass concentrations as MLVSS correspond well with the predicted model values for cell growth. The values of the kinetic parameters obtained were as follows: K_s = 183 mg/L, Y= 0.21 mg biomass/mg NO_3 and $\mu_m = 0.136\ hr^{-1}$.

Kurt et al. [8] studied autotrophic hydrogen-oxidizing denitrifying bacteria using a bench-scale fluidized bed reactor. The process was modeled using a double Monod-saturation function. Nitrate and hydrogen were assumed to be limiting substrates in the first denitrification step, and nitrite and hydrogen were assumed limiting in the second step. The denitrification rate was shown to be more strongly dependent on the nitrate concentration than on hydrogen. Hydrogen is only slightly soluble in water (1.6 mg/L @ 20°C), and the Monod half-velocity constant for hydrogen was found to be less than 1% of saturation. Nitrite tended to accumulate in batch tests, but complete denitrification occurred in continuous-flow experiments with sufficiently long residence times.

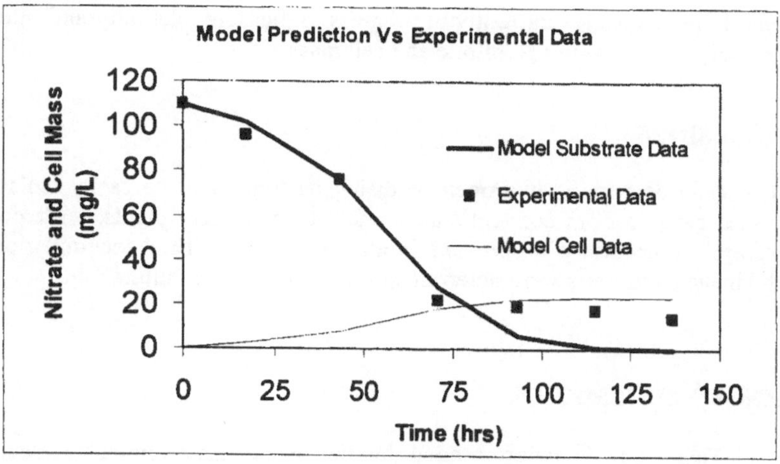

Figure 3: Modeling Results for Batch Experiments

The most challenging operating parameter was pH control. All three reactors required pH control to maintain an average pH of about 7.2. The measured pH in the reactors before pH control is presented in Figure 4.

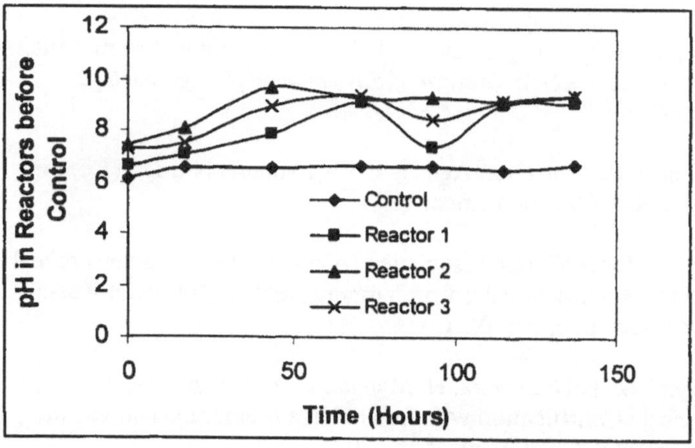

Figure 4: pH in Batch Reactors before pH Control

The pH data indicates that pH values approached 9.0 pH units during the experiment duration. This indicates that nitrite accumulations may occur in the system. Experiments are currently in progress to measure alkalinity and nitrite concentrations along with pH, nitrate and cell mass.

CONCLUSIONS

Autotrophic dentrifying hydrogen oxidising bacteria acan be easily isolated from seed bacteria from conventional wastewater treatment plants. Adequate pH control is necessary to prevent inhibition due to nitrite accumulations. Batch kinetic parameters were determined for the developed culture

ACKNOWLEDGEMENTS

This study is being funded by the Water Environment Research Foundation located in Alexandria, Virginia. However, this study has not been subjected to the Foundation's peer and administrative review and therefore may not necessarily reflect the views of the Foundation, and no official endorsement should be inferred.

REFERENCES

1. Kapoor, A. and Viraraghavan, T., 1997, Nitrate removal from drinking water – Review, *ASCE Journal of Environmental Engineering*, 123:4:371-380.

2. Environmental Protection Agency (EPA), Manual Nitrogen Control, *EPA/625/R-93/010*, September 1993.

3. American Water Works Association (AWWA), 1995, Surveys reveal pesticides, nitrates in water from farming areas, *J. American Water Works Association*, 87(10), 13-14.

4. Delanghe, B., F. Nakamura, H. Myoga and Y. Magara, 1994, Biological Denitrification with Ethanol in a Membrane Bioreactor, *Environmental Technology*, 15, 61-70.

5. Lazarova, V.Z., B. Capdeville and L. Nikolov, 1992, Biofilm Performance of a Fluidized Bed Biofilm Reactor for Drinking Water Denitrification, *Water Science Technology*, 26(3-4), 555-566.

6. Metcalf and Eddy, Inc., 1991, Wastewater Engineering: Treatment, Disposal, and Reuse, *McGraw Hill*, 3rd edition, New York, NY.

7. Gross, H. and K. Treutler, 1986, Biological Denitrification Process with Hydrogen Oxidizing Bacteria for Drinking Water Treatment, *Aqua*, (5), 288-290.

8. Kurt, M., Dunn, J. and Bourne, J. R., 1987, Biological denitrification of drinking water using autotrophic organisms with hydrogen in a fluidized-bed reactor. *Biotechnology and Bioengineering*, 29:493-501.

9. Claus, G., and Kutzner, H. J., 1985, Autotrophic Denitrification Thiobacillus denitrificans, *Appl. Microbiology and Biotechnology*, 22:289-296.

10. Barrenstein, A., Kramer, U., Obermann, P, 1986, Underground treatment of nitrate rich groundwater by infiltration with treated wastewater or methane-rich natural gas, *DVGW-Schriftenreihe*, Wasser, Frankfurt, West Germany, 106.

11. Gross, H. and K. Treutler, 1986, Biological Denitrification Process with Hydrogen Oxidizing Bacteria for Drinking Water Treatment, *Aqua*, (5), 288-290.

12. Ginocchio, J., 1983, Process and apparatus for the denitrification of groundwater, *European Patent Appl., EP 86863 Al*, Aug. 31.

13. Gantzer, C. J., 1995, Membrane Dissolution of Hydrogen for Biological Nitrate Removal, *Proceedings of Water Env. Fed. 8th Annual Conference*, Miami Beach, FL.

14. AWWA, APHA and WEF, 1995 *Standard Methods for the Evaluation of Waste Water*, Washington, D.C.

15. Monod, J. 1949, The Growth of Bacterial Cultures, *Ann. Rev. Microbiol.*, Vol. 3.

DESIGN AND CONSTRUCTION OF A FIELD-SCALED PHOTO-CHEMICAL REMEDIATION REACTOR

KENNETH Y. LEE
Department of Civil & Environmental Engineering
Rutgers, The State University of New Jersey, Piscataway, NJ 08854

JOHANNES KHINAST
Department of Chemical and Biochemical Engineering
Rutgers, The State University of New Jersey, Piscataway, NJ 08854

JEFF J-Y. LEE
Department of Civil & Environmental Engineering
Rutgers, The State University of New Jersey, Piscataway, NJ 08854

JOSEPH R. STENCEL
Department of Civil & Environmental Engineering
Rutgers, The State University of New Jersey, Piscataway, NJ 08854

ABSTRACT

In this study, a field-scale photo-chemical remediation reactor is designed and constructed. The reactor is used for destruction of vapor-phase volatile organic compounds (VOC), particularly chlorinated hydrocarbons, based on ultra violet (UV) photo-chemical reactions. UV light, when emitted at an effective absorption frequency, will cleave a VOC's carbon-chlorine bond, initiating long-chain radical reactions, thus reducing harmful contaminants to harmless products. The reactor consists of a stainless steel tubular vessel with internal dimensions 31.9 cm in diameter and 1.05 m in length. Three stainless steel baffles are welded inside the reactor to create a well-mixed vapor phase and homogeneous UV contact time. Sixteen Suprasil glass sleeves (Heraeus, Inc.) with external dimensions of 25 mm in diameter and 1.1 m in length are inserted along the length of the reactor positioned in a geometrical pattern to provide maximum UV exposure. The Suprasil glass sleeves are attached and sealed to the stainless steel reactor using a "stuffing-box" compression fitting. An amalgam UV lamp (Heraeus, Inc. NIQ 200/110) is placed inside every Suprasil glass sleeve. The Suprasil glass sleeves provide barrier protection between the UV lamps and the vapor-phase contaminant, and at the same time allow transmission of UV light at the primary wavelengths of 185 and 254 nm to the interior of the reactor. This design allows easy alteration and maintenance of UV lamps. The reactor is designed to operate with a vapor-phase temperature range from ambient to up to 300 degrees Celsius. A gas heater is installed to increase the influent vapor phase temperature. The reactor is mounted on a mobile frame to allow transport to field sites. The technology can be effectively applied to gaseous or liquid waste streams containing chlorocarbons. The next phase of the study will be to examine the destruction efficiency of various VOCs. Contaminants of interest include tetrachloroethylene (PCE) and trichloroethylene (TCE).

INTRODUCTION

The design of the reactor described is based on extensive bench-scale studies [1, 2] performed at Energia, Inc. (Princeton, NJ), which indicated that the use of photo-chemical remediation (PCR) technology should be a "safe and cost-effective destruction of hazardous chlorine-containing solvent." The PCR technology is based on a synergistic effect obtained when UV photo-initiation is combined with either a reducing gas (e.g., hydrogen) or oxidative atmosphere (ambient air) or combination of both. The combination promotes long-chain chemical reactions, which rapidly and efficiently destroy halocarbons. Our initial experiments with this reactor will be without the reducing gas.

The University of Dayton Research Institute (UDRI) had proposed a four chamber "photothermal detoxificxation unit (PDU)" which did not proceed beyond a bench-scale test [3]. The theory for the photothermal destruction is discussed in a paper by *Chen et al.* [4]. The design and construction of the PCR reactor is part of a pilot-scale demonstration project for Energia, Inc. It increases the size of the bench-top reactor, adds baffles for better gas mixing and longer residence time, increases the number and size of UV lamps (up to 16 one-meter lamps) to accommodate the larger reactor, and puts the heating in-line with the gas flow. While the process is exothermic, the heating option allows the heating of input gas, if required, for increasing the reaction rates. All the hardware is mounted on a mobile frame for technology demonstration at a field site. Like the UDRI design [3] this design uses multi-lamps in a cylindrical chamber. This design is modified, however, to improve the uniformity of the emitted light absorbed by the reactants. The main flow in the reaction chamber is longitudinal (i.e., parallel to the lamps) rather than transverse as in the UDRI proposal. This flow configuration is expected to

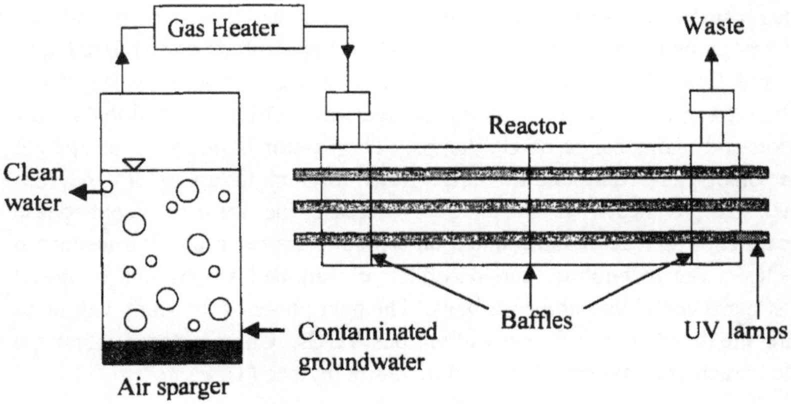

Figure 1. Experimental setup configuration.

provide better uniformity. Multi-chamber configurations can also be used and connected in series, parallel, or some type of combination of both.

The Figure 1 schematic shows the vapor flow path of the presently designed reactor. The system includes an air-sparger unit; a Chromalox 4kW gas heater, which will help sustain radical chain-reactions; a photo-chemical reaction chamber with baffles to increase mixing through this vessel; and up to 16-one meter length UV-lamps.

REACTOR VESSEL

Table 1 shows the design parameters of the reactor. Figure 2 shows some materials that were considered for the reactor vessel because of their low coefficients of thermal expansion (CTE). The material selected for the reactor vessel was 304 stainless steel because of cost considerations, ability to resist corrosion, and ease in fabrication.

Table 1. Reactor design parameters.

Operating temperature:	Ambient – 300 °C
Operating pressure:	1 atm
Gas flowrate:	1-5 L/s
Ports:	5 total (3 for sampling, 1 inlet and 1 outlet)
Number of UV lamps:	Variable, maximum 16
Reactor material:	Stainless steel

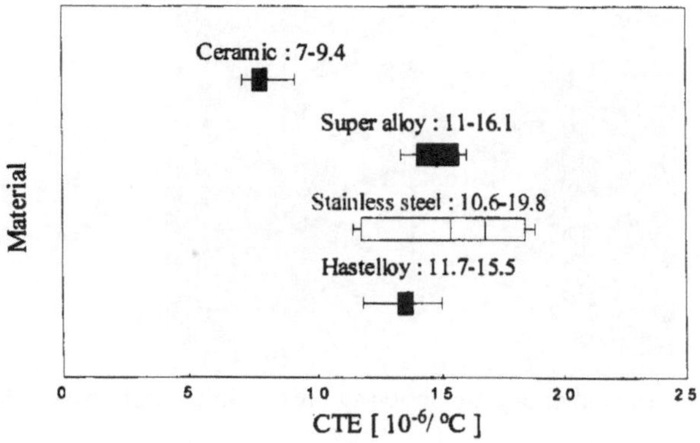

Figure 2. Coefficient of Thermal Expansion (CTE) of select materials.

The reactor has an inner diameter of 31.9 cm (12.57 inch) and a length of 1.05 m (41.3 inch). Figure 3 shows detailed dimensions of the reactor. The interior volume without any hardware is approximately 84.1 L and this volume is reduced to approximately 73.7 L with the insertion of the Suprasil glass sleeves holding the UV lamps and internal baffles. The device contains an influent and effluent port and three sampling ports as part of the reactor design. The three interior baffles within the reactor vessel help to provide a turbulent flow of gas through the reactor. Figure 4 shows a close-up picture of the baffles. The one-inch stainless steel end plate flanges are mounted to the reactor chamber using a 1/32-inch thick graphite gasket and 24, 0.5-inch by 2.5-inch length bolts.

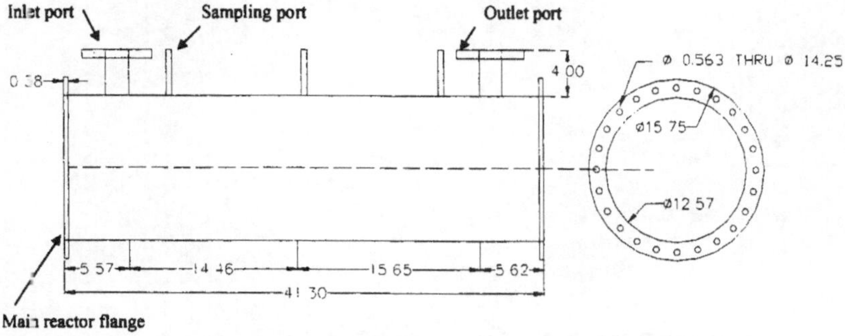

Figure 3. Reactor vessel dimensions. Units shown in inches.

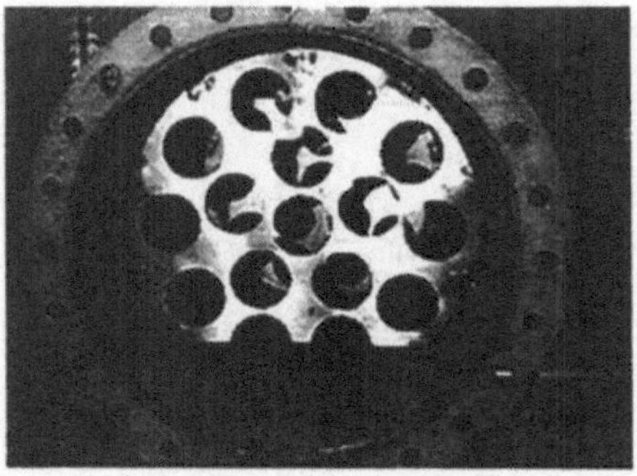

Figure 4. Removed end-plate showing baffles in reactor vessel.

The end plate flanges allow for the passage of the Suprasil glass sleeves. It should be noted that the Suprasil glass sleeves protect the UV lamps from contaminant vapor while allowing transmission of UV light into the reactor. The 16 Suprasil glass sleeves, and thus UV-lamps, are geometrical positioned by the end-plate flange as shown in the reactor photo Figure 5. Two 1/4-inch thick graphite compression seals manufactured by Garlock Sealing Technologies are compressed with an aluminum gland piece (Figure 6 and 7) and held in place by three 1-1/4 inch 10/32 socket set screws which forms a "stuffing box" type seal between each Suprasil glass sleeve and stainless end plate flange.

Figure 5. Reactor showing geometric configuration of lamps through the end-plate flange.

Figure 6. Side and top view of "stuffing-box" compression gland.

UV LAMPS

Low-pressure mercury lamps have two resonance lines, namely, 185 and 254 nm. The lamps presently used are manufactured by Heraeus of Germany (Heraeus, Inc., NIQ 200/110). Input power is 200 W per lamp. Their length is 106.7 cm, with a diameter of 22 mm, and with a UV output of ~80 W (~40%). The Suprasil glass sleeves have a 25 mm inner diameter and 28 mm outer diameter with a length of 110 cm. Each lamp has its own ballast and starter for ease of design and allows for various configurations to be used in our optimizing of the UV irradiation during experiments. The UV lamps are placed inside of the Suprasil glass sleeve using O-rings (Figure 7) to keep the

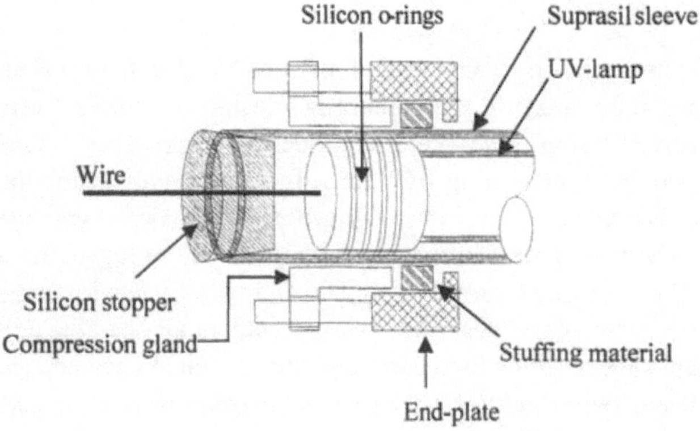

Figure 7. "Stuffing-box" compression fitting. This figure also shows o-rings around UV lamp to keep them off the Suprasil glass sleeve.

lamps off of the sleeve surface. The area between the lamps and sleeve is purged with nitrogen during placement to minimize the amount of ozone produced within this space. A large silicon stopper is placed at each end of the Suprasil glass sleeve to prevent nitrogen leaking.

WASTE APPLICABILITY

The PCR reactor has commercial potential for the following remedial operations:

1. In-situ treatment of chlorinated effluents discharged from soil venting extraction (SVE).

2. Direct treatment of off-gas streams containing chlorocarbons.

3. Treatment of groundwater contaminated with chlorocarbons in tandem with air-sparging.

4. Pretreatment of waste streams entering catalytic oxidation systems.

5. On-line remediation in R&D and testing laboratories, chemical hoods, clean rooms, depainting booths, etc.

The PCR reactor can operate as a stand-alone system or in combination with other remedial systems.

SUMMARY

This paper focuses on the design and construction of a field-scaled, photo-chemical remediation reactor. Numerous design obstacles were encountered and resolved using innovative engineering approaches. One particular difficulty was gastight sealing of Suprasil glass sleeves inside the stainless steel reactor. For this problem, a "stuffing-box" compression graphite fitting was selected. The use of stainless steel allows for a more rugged field device although the Suprasil glass sleeves and UV lamps still require some careful handling. The reactor vessel described is used with an air-sparging unit to produce contaminant vapors which then can be photo-chemically remediated to a level which will not require either air emission permits or costly waste disposal. The next phase of the study will be to examine the destruction efficiency of various VOCs. Contaminants of interest include tetrachloroethylene (PCE) and trichloroethylene (TCE).

REFERENCES

1. EPA, U.S., 1994. Emerging technology bulletin, reductive photo-dechlorination (RPD) process for safe conversion of hazardous chlorocarbon waste streams, Energia, Inc. of Princeton, NJ, EPA/540/F-94/508, Cincinnati, OH.

2. EPA, U.S., 1999. Superfund innovative technology evaluation (SITE) program: Emerging technology program, Vol. 2, pp. 38, 122, EPA/540/R-99/500, Cincinnati, OH.

3. EPA, U.S., 1995. Superfund innovative technology evaluation (SITE) program: Development of photothermal detoxification unit, EPA/540/F-95/505, Cincinnati, OH.

4. Chen, C. T., Graham, J. L., and B. Dellinger, 1995. Photothermal destruction of the vapor of organic compounds, Waste Management 15(2), 159-170.

ACKNOWLEDGEMENTS

The authors would like to thank Energia, Inc. for funding of this project, and to Drs. Moshe Lavid and Moisey Teytelboym for technical support. Jeff J-Y. Lee is also grateful for a graduate scholarship from New Jersey Water Resources Research Institute (NJWRRI).

MODELING THE FEASIBILITY OF ZERO LIQUID DISCHARGE SOLUTIONS IN THE PROCESS INDUSTRY

BRIAN J. DERICKS, MARIANO J. SAVELSKI
Rowan University
Chemical Engineering
201 Mullica Hill Rd.
Glassboro, NJ 08028

ANANTHA P. R. KOPPOL, MIGUEL J. BAGAJEWICZ
University of Oklahoma
Chemical Engineering and Materials Science
100 E. Boyd St.
Norman, OK 73019

ABSTRACT

In this work, the issue of minimizing freshwater usage is addressed by determining technical and economical feasibility of zero liquid discharge options in different industries. This approach uses mathematical programming to automatically generate the optimal solution to achieve zero liquid discharge.

INTRODUCTION

Water is required in several fundamental process operation units such as scrubbing, acting as extraction agent, generation of steam, etc. The problem has received attention from several researchers. Even though the problem was suggested by Prof. Umeda in 1980 [1], the field owes its formal existence because of the pioneering work of El Halwagi and Manousiouthakis [2], in mass exchange networks, seminal contributions that were picked up by Wang and Smith [3] and applied them to the particulars of this problem. The list of subsequent work is extensive and can be obtained from a recent review [4].

Wang and Smith [5] also introduced the use of regeneration, but they concentrated on systems without recycles of water. Zero discharge, therefore, was excluded. Bagajewicz and Savelski [6] presented targeting models for overall freshwater minimization, which include regeneration with and without recycles. Results from this paper show that zero liquid discharge is feasible only if regeneration has an outlet concentration that is sufficiently small. Otherwise, recycles might exist, but liquid discharge is not completely eliminated. In addition, after these targeting models are solved, several alternatives can be explored to minimize investment cost because the problem has many solutions with the same consumption. This paper builds on these developments and studies zero discharge cycles. In this paper, the targeting models for this application are reviewed and two case studies are analyzed: water minimization in the production of tricresyl phosphate and in a paper mill process.

PROBLEM STATEMENT

Given a set of processes, the problem is to determine an arrangement of connections from process to process that will reduce the consumption of freshwater, possibly eliminating it, while receiving water of adequate quality. In the Water/Wastewater Allocation Planning (WAP) problem limits on inlet and outlet concentration of contaminants are set, in the same way as posed by Wang and Smith [3]. The outlet concentration limits account for corrosion, fouling, maximum solubility, etc, while the inlet is set to limit the total flowrate through processes. Bagajewicz and Savelski [6] exploit the necessary conditions of optimality, given in [4], to obtain the following linear model.

$$\left.\begin{aligned}
& P1 = Min \sum_j F_j^w \\
& s.t. \\
& F_j^w + \sum_{i\in P_j} F_{i,j} - \sum_{k\in R_j} F_{j,k} - F_{j,out} = 0 && \forall j \\
& \sum_{i\in P_j} F_{i,j}\left(C_{i,out}^{max} - C_{j,in}^{max}\right) - F_j^w C_{j,in}^{max} \le 0 && \forall j \\
& \sum_{i\in P_j} F_{i,j}\left(C_{i,out}^{max} - C_{j,out}^{max}\right) - F_j^w C_{j,out}^{max} + L_j = 0 && \forall j
\end{aligned}\right\} \quad (1)$$

This single component model is rigorous and has no approximations other than the assumptions of constant load and maximum inlet and outlet concentrations. Once this model has produced a target, the objective function can be changed to minimize the number of connections or their linear combination, which reflect the piping cost. An additional constraint is also added limiting the freshwater consumption to the targeted value.

WAP Model with Regeneration

Bagajewicz and Savelski [6] added a regeneration unit to the above model and proved linearity is still possible. They also proved that the outlet concentration of the regeneration unit should be at its minimum. The model is similar to P1, which provides a target of freshwater (α). Because multiple solutions featuring the same freshwater consumption are possible, a second phase, minimizing the total cost of regeneration is added.

$$P2 = Min\{Regeneration\ Cost\} \\ s.t. \\ \sum_j F_j^w = \alpha \\ Mass\ and\ component\ balance \quad (2)$$

where, the regeneration cost is directly related to the total flowrate ($a_R \sum_j F_{j,T}$) or to the total load removed ($a'_R \sum_j F_{j,T}(C_{j,out}^{max} - C_T)$) depending on the case.

The above problem allows recycles and therefore is suitable to identify zero discharge solutions. Nevertheless, Bagajewicz and Savelski [6], proposed additional constraints that allow the identification of solutions without recycles, which in this case are of no importance.

In the third phase, the capital cost is minimized using the model proposed by Bagajewicz and Savelski [6]. This is done by minimizing the total number of inlet, outlet and interconnections between the processes using the targets for freshwater (α) and regeneration cost (β) provided by P1 and P2 as follows:

$$\left.\begin{array}{ll} P3 = Min \left\{ \sum_{i,j} Y_{i,j} + \sum_{w,j} Y_{w,j} + \sum_{j,o} Y_{j,o} \right\} & \\ s.t. & \\ \sum_j F_j^w = \alpha & \\ Regeneration\ Cost = \beta & \\ Mass\ and\ component\ balances & \\ F_{i,j} - U\, Y_{i,j} \le 0 & \forall i, \forall j \\ F_j^w - U\, Y_{w,j} \le 0 & \forall j \\ F_{j,out} - U\, Y_{j,o} \le 0 & \forall j \\ Y_{i,j}, Y_{w,j}, Y_{j,o} \in \{0,1\} & \end{array}\right\} \quad (3)$$

In this paper, the above three models are used to obtain the networks. In addition to regeneration cost, the operating cost includes cost of freshwater and its final treatment given by $a_w \sum_j F_j^W + a_T \sum_j F_{j,out}$.

CASE STUDIES

Three industrial cases are presented: The production of tricresyl phosphate and ethyl chloride, from El-Halwagi [7], and a paper mill (Tripathi [8]; Brezniak[9]; Smook [10]).

Case #1: Tricresyl Phosphate

TABLE I: DATA FOR CASE #1

Process Number	Process Description	Load (g/h)	C_{in}^{max} (ppm)	C_{out}^{max} (ppm)
1	Washing I	626.22	5.00	76.00
2	Washing II	0.6174	0.00	0.07
3	Scrubber I	1152.1	30.00	411.00
4	Scrubber II	246.24	30.00	144.00
5	Flare Seal Pot	163.35	100.00	281.50

In this process, tricresyl phosphate is being purified to remove all of the unreacted cresol. The process incorporates two washers for product purification, two scrubbers and a flare seal pot for the removal of contaminant from off-gas. Table I presents the data for the process. One can use light gas treatment or oil treatment as a regeneration step, which can clean up to 3 ppm or 40 ppm, respectively. Costs are: a_w= 1.5 *\$/Ton*, a_T= 1.0 *\$/Ton*. The cost of gas and oil regeneration are dependent on load, a'_R= 434.7 \$/Kg and a'_R= 116.6 \$/Kg for gas treatment and oil treatment, respectively (El-Halwagi [7]). Using this data, P1, P2 and P3 are solved to obtain the networks with and without regeneration as shown in Figures 1A and 1B, where process 6 is regeneration unit with outlet concentration of 3ppm.

FIGURE 1A: SOLUTION NETWORK WITHOUT REGENERATION

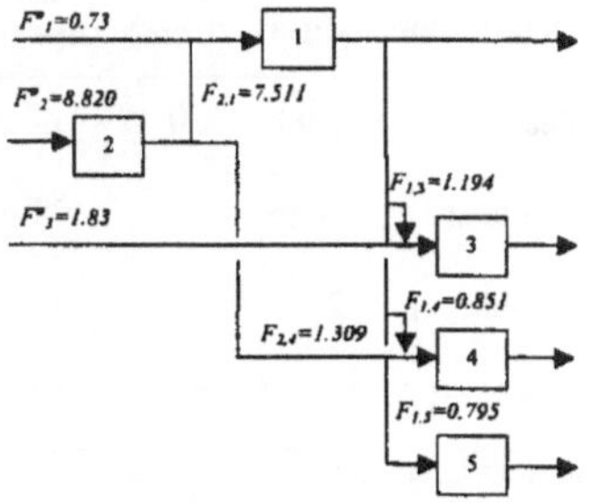

FIGURE 1B: SOLUTION NETWORK WITH REGENERATION

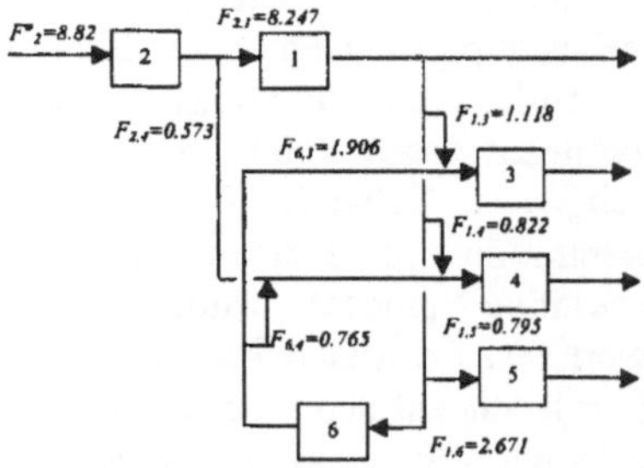

The freshwater saving without regeneration after optimizing is 52%, which is 12.34 te/h freshwater. Thus, reducing the cost from \$498,120 per year to \$239,085 per year. Addition of treatment further reduces freshwater, at the same time increases the cost. Figures 2 and 3 depict cost and freshwater flowrate variation with outlet concentration of the regeneration.

As seen in Figure 3, zero discharge is not possible as the minimum outlet concentration for the light gas treatment is 3 ppm. This is because process 2 requires freshwater (0 ppm). For an outlet of 3 ppm, the cost increases to \$897,331 per year, because light gas treatment is more expensive than cost of freshwater and its final treatment. Thus, oil treatment becomes a cheaper option. Oil treatment reduces freshwater consumption to 9.03 te/hr, with a cost of

FIGURE 2: COST EVALUATION FOR TREATMENT PROCESS

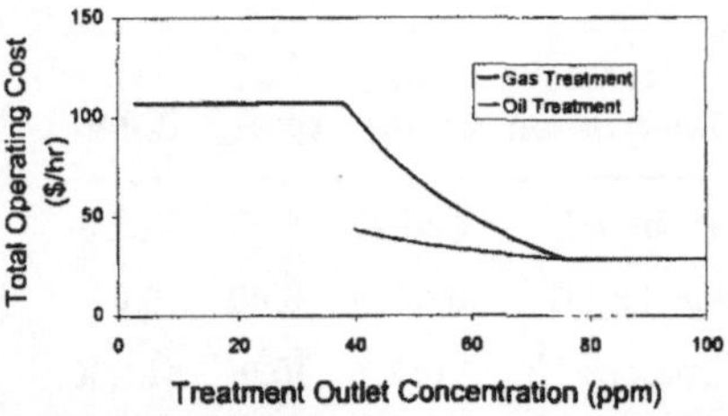

FIGURE 3 : FRESHWATER REDUCTION THROUGH TREATMENT PROCESS

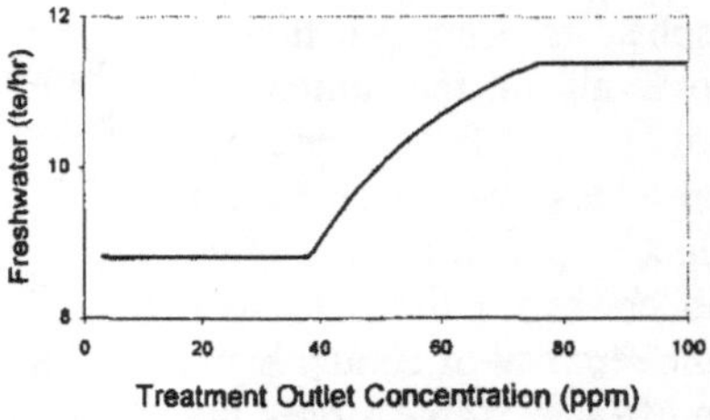

$365,054 per year. As pointed out above, zero discharge is not possible in this case.

Case #2: Ethyl Chloride

Ethyl chloride is produced by catalytically reacting ethanol and hydrochloric acid. The following process uses ethylene as a raw material to produce ethanol through a catalytic hydration process in Reactor I. In this process no contaminant is produced so the water may be reused. The ethanol is then sent to Reactor II where it is reacted with hydrochloric acid to produce ethyl chloride. The effluent from Reactor II is sent through two scrubbers to remove the contaminant of our focus, chloroethanol, and unreacted raw materials to purify the product. Inlet and outlet concentrations and contaminant loads upon processes are shown in Table II. The outlet concentration of Reactor II, a regenerative process within the network, was considered a variable that is dependent upon the reactor's inlet concentration by El-Halwagi [7]. This concentration can be treated as a parameter, so that a linear formulation can be used. For space reasons we use a fixed value.

TABLE II: DATA FOR CASE #2

Process Number	Process Description	Load (Kg/h)	C_{in}^{max} (ppm)	C_{out}^{max} (ppm)
1	Reactor I	0.00	0	0
2	Scrubber I	29.16	8	98
3	Scrubber II	2.75	0	9
4	Reactor II	21.33^R	49.5	10

R – Regenerated Load (Kg/h)

Air or zeolite treatment can be used as a regeneration step, each can clean up to 10 ppm or 2 ppm of chloroethanol, respectively. The cost of air and zeolite regeneration are dependent on load, a'_R= 5,000 $/Kg

FIGURE 4: ETHYL CHLORIDE SOLUTION NETWORK WITHOUT REGENERATION

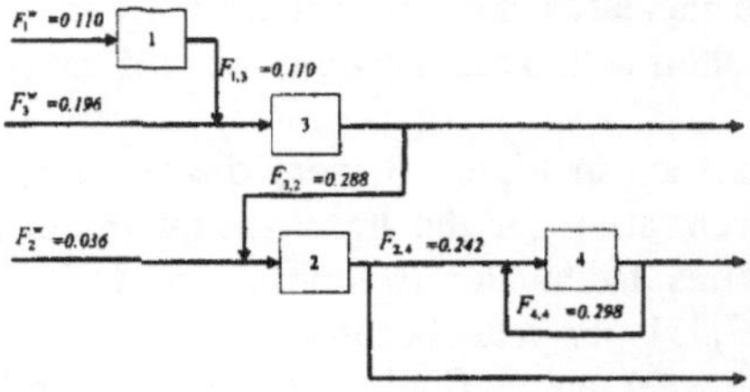

FIGURE 5: FRESHWATER SAVINGS THROUGH ADDITIONAL TREATMENT

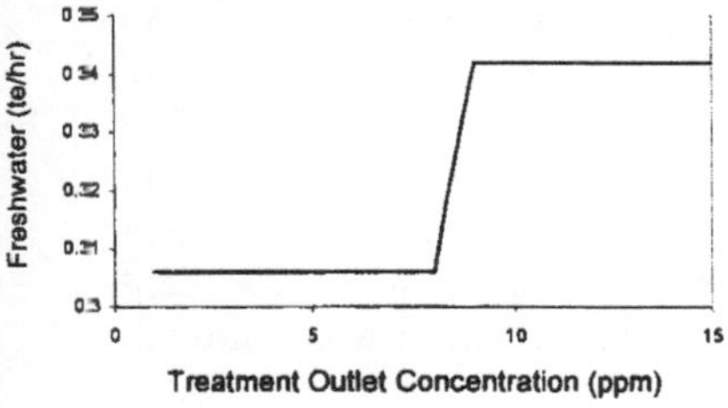

FIGURE 6: SOLUTION NETWORK WITH ZEOLITE TREATMENT

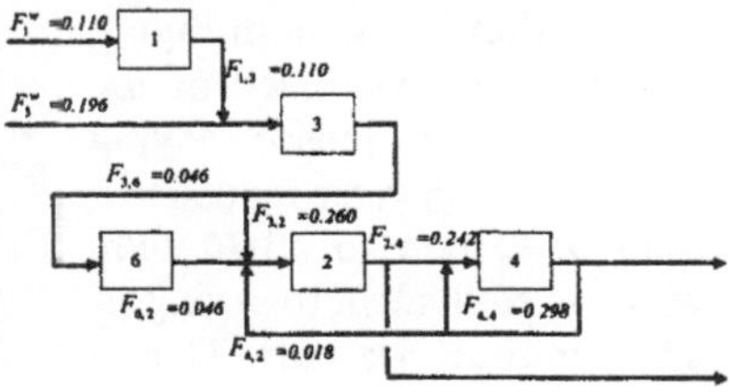

and a'_R= 58,333 \$/Kg for gas treatment and oil treatment, respectively [7]. Other costs are: a_w = 1.5 *\$/te*, a_T= 1.0 *\$/te.*

By optimizing the network without any regeneration, it is found that the freshwater consumption is reduced from 1.28 te/hr to 0.34 te/hr. This network, shown in Figure 4, reduces the cost of water from \$26,880 per year to \$7,190 per year.

By utilizing treatment options, the freshwater usage may be reduced depending upon the outlet concentration of the treatment. The dependence is shown in Figure 5. At 9 ppm, the freshwater usage drops to 0.306 te/hr. The abrupt change in water usage is due to the availability of process 3 to connect to the treatment process optimally. Since a treatment to reduce freshwater consumption requires an outlet concentration of 9 ppm, no treatment is needed with an outlet concentration above 9 ppm. The network utilizing zeolite treatment, in Figure 6, reduces the freshwater usage to 0.306 te/hr with a new total operating cost of \$174,005 per year. The strict constraints on the inlet concentration of process 1 and 4, and the inability for the treatment processes to attain 0 ppm of contaminant do not allow a zero liquid discharge solution. Further research into different treatment technologies and methods to reduce the load of contaminant on the processes are suggested.

Case #3: Paper Mill

Paper mills today are capable of achieving zero discharge as shown by the following model using physical and membrane treatments. Recycling the whitewater (paper-rich wastewater) allows the process to reuse water and recover lost paper. In this case, a 600 te/day non-bleach paperboard mill was modeled and optimized. The mill consists of chests for pulping and diluting paper fibers, paper machine, showers, and added treatments. The treatments of concern are physical and membrane treatment. In addition, one should consider the paper machine as two streams (Rich and Lean), which have been modeled as additional regeneration processes with fixed maximum outlet and inlet concentrations, and fixed load as shown in Table III. An additional constraint for the paper mill model is that all the water from process 1 is used in paper machine (Rich and Lean), to prevent paper loss through other processes. The physical and membrane treatments have a minimum outlet concentration of 30

TABLE III: DATA FOR CASE #3

Process Number	Process Description	Load (Kg/h)	C_{in}^{max} (ppm)	C_{out}^{max} (ppm)
1	Pulping / Dilution	24300	500	5000
2	Deckle Showers	21.67	100	500
3	Cylinder Showers	37.50	300	600
4	Felt Showers	3.33	20	100
5	Paper Machine (Rich)	18225^R	5000	500
6	Paper Machine (Lean)	6480^R	5000	200

R – Regenerated Load (Kg/h)

and 2 ppm, respectively. The cost for freshwater and wastewater treatment is 1.65 \$/te of water, and $a_{R,Phys}$ = 0.15 \$/te, $a_{R,\ Memb}$ = 0.9 \$/te (Wiseman *et al.* [11] and Lagasé et al. [12]). Here regeneration cost is dependent on the amount of water through it.

The optimized network corresponding to reuse without regeneration is shown in Figure 7. It is observed, much of the white-water is reused saving 99.2% of freshwater reducing it from 5099 te/h to 43.74 te/h, which is equivalent to saving of \$77.3 million per year. This is why reusing water in a paper mill is so vital, which saves both water and paper.

The freshwater intake is reduced even further using regeneration. The optimized networks including regeneration are shown in Figure 8 and 9, where regeneration units are depicted as process 7 for physical and process 8 for membrane. The outlet concentrations of regeneration in these networks are 30 and 2 ppm, respectively.

Using physical treatment reduces freshwater consumption to 13.88 te/hr, which results in an operating cost of \$236,573 per year. This cost is strongly dependent upon the outlet concentration of the treatment. Figures 10 and 11 depict the cost and the freshwater variation with the outlet concentration of treatment. These figures reveal that if the outlet concentration of physical treatment can be reduced to 20 ppm a zero discharge system is possible, and the cost would further reduce!

FIGURE 7: PAPER MILL SOLUTION NETWORK WITHOUT REGENERATION

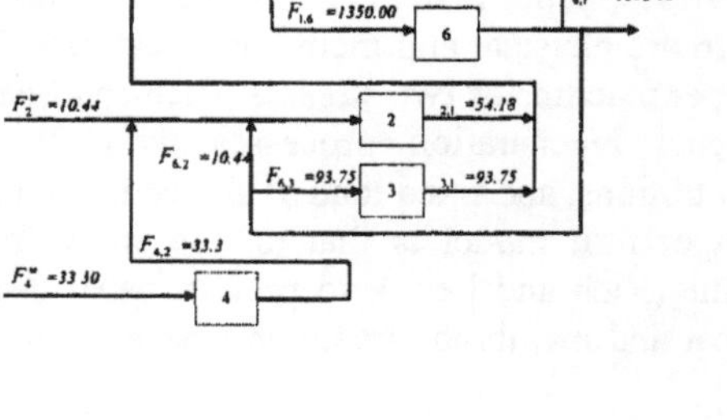

By adding a membrane treatment consisting of ultrafiltration and reverse osmosis to the physical treatment, additional freshwater savings are obtained. Zero discharge system occurs when the membrane system reduces the concentration to 20 ppm, which is higher than the originally assumed 2 ppm. However, cost continues to drop beyond 20 ppm. This results in a yearly operating cost of $336,314. This is the cost of regeneration alone, but if effluent release is an issue, it is necessary.

FIGURE 8: SOLUTION NETWORK WITH PHYSICAL TREATMENT

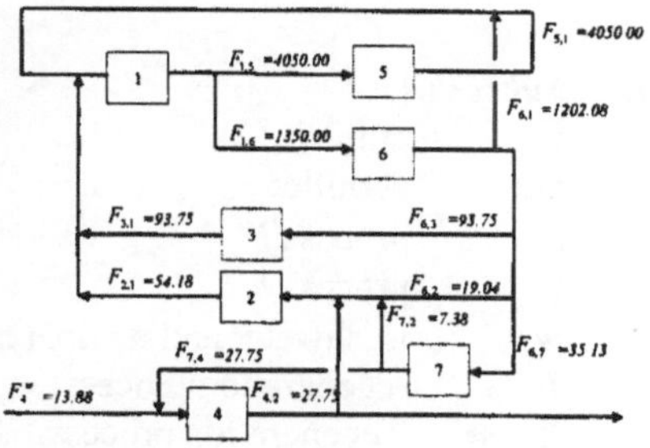

FIGURE 9: SOLUTION NETWORK WITH MEMBRANE TREATMENT

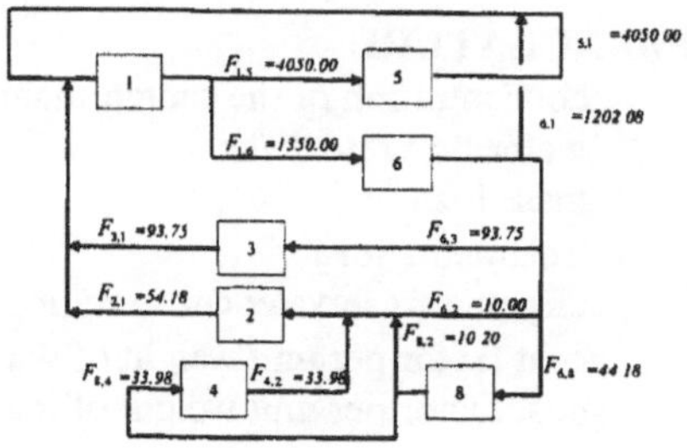

FIGURE 10: OPERATING COST EVALUATION FOR ADDITIONAL TREATMENT PROCESS

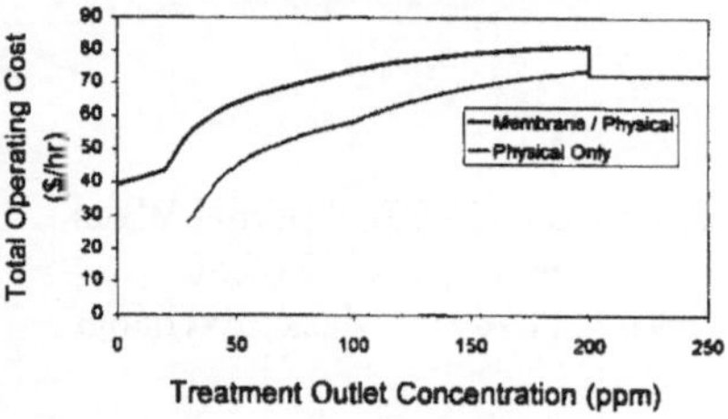

FIGURE 11: FRESHWATER REDUCTION THROUGH TREATMENT PROCESS

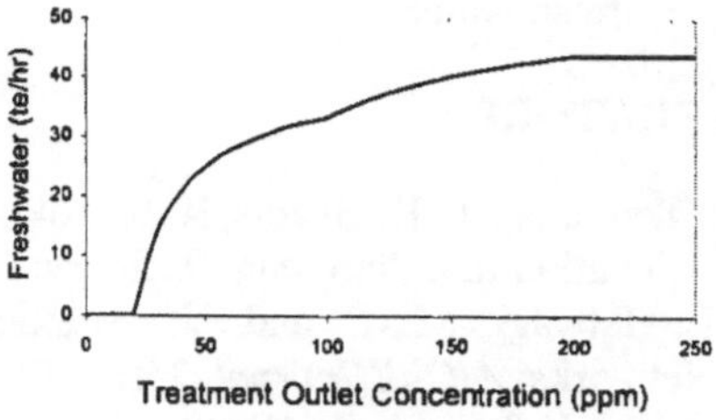

CONCLUSION

In this paper, zero discharge solutions for systems with one contaminant are explored using a linear programming model recently developed. The relationship between the cost of regeneration and the cost of freshwater as well as discharge treatment is shown to be the determining factor in the structure and economical feasibility of zero or partial liquid discharge cycles.

NOMENCLATURE

C	concentration of the contaminant, ppm
F	water flowrate, te/hr
L_j	mass load on j
T_j	treatments for j
α	targeted freshwater consumption
a	cost factor per unit weight of water
a'	cost factor per unit weight of load

Subscripts

in	at inlet
out	at outlet
j	process j
i	process i
w	freshwater and its final treatm
T	regeneration process
R	Regeneration process

Superscripts

min	Minimum
max	Maximum
w	fresh water

LITERATURE

1. Takama N., T. Kuriyama, K. Shiroko and T. Umeda. (1980). Optimal Water Allocation in a Petroleum Refinery. *Comp. & Chem. Eng.*, 4, 251-258.
2. El-Halwagi M.M. and V. Manousiouthakis. (1989). Mass Exchanger Networks. *AIChE Journal*, 35(8), 1233.
3. Wang Y.P. and R. Smith. (1994). Wastewater Minimization. *Chemical Engineering Science*, 49(7), 981.
4. Bagajewicz M. and M. Savelski. (2000). On the Optimality Conditions of Water Utilization Systems in Process Plants with Single Contaminants. *Chemical Engineering Science*, 55, 5035-5048.
5. Wang Y.P. and R. Smith. (1994b). Design of Distributed Effluent Treatment Systems. *Chemical Engineering Science*, 49(18), 3127.
6. Bagajewicz M. and M. Savelski. To appear (2001). On The Use of Linear Models for the Design of Water Utilization Systems in Refineries and Process Plants. *Chemical Engineering Research and Design.*
7. El-Halwagi M. M. (1997). *Pollution Prevention Through Process Integration.* San Diego, California: Academic Press.
8. Tripathi, P. (1996). Pinch Technology Reduces Wastewater: Mass Exchange Integration Maximizes Water Recycling at a Paper Mill. *Chemical Engineering*, Nov., 87-90.
9. Brezniak, S. (1999). *Process Water Quality and Water Reuse Practices at Low- and Zero Discharge Recycled Paperboard Mills.* Research Triangle Park, NC: National Council for Air and Stream Improvement.

10. Smook G. A. (1994). *Handbook for Pulp and Paper Technologists, 2nd Ed.* Vancouver, B.C.: Angus Wilde Publications Inc.
11. Wiseman, N. and Ogden, G. (1996). Zero Liquid Effluent Technologies for the Paper Industry. *Paper Technology*, 37, 31-38.
12. Lagasé, Miner, and Stuart, P. R. (1998). Cost Associated with Implementation of Zero Effluent Discharge at Recycled Fiber Paperboard Mills. *TAPPI Proceedings*, 3, 1011-1018.

GIS Visualization of mercury concentrations in sediments of the New York/New Jersey Harbor

Alexander F. Gregory[1] and Yuri Gorokhovich[2]

[1]Earth and Environmental Engineering Department
Columbia University

[2]NYC, DEP

The study has compiled a database of surficial sediment mercury concentration samples with 469 unique values over a period of 10 years (1988 2000). Preliminary results of the Harbor GIS study are shown below. Data have been presented for two continuous years in the early 1990s (1991 & 1992) from two different studies. Black dots in each figure represent mercury surface sediment concentration greater than 2 ppm. In each figure, high zones of contamination appear to be mostly within Arthur Kill, East River, and Newark Bay. However, the number the number of sediment samples with concentrations greater than 2 ppm decreased dramatically from one year to the next. For the 1991 study (NOAA 1995), the highest concentration of 15 ppm was measured in Arthur Kill near the Island of Meadows. For the 1992 (NYC DEP 1993), the highest concentration of 7 ppm was measured in Newark Bay off Shooter's Island.

Integrated Waste Management Options for New York City Municipal Solid Waste

ClaireTodd and Nickolas.J. Themelis

Earth and Environmental Engineering Department
Henry Krumb School of Mines
Columbia University in the City of New York
New York, NY 10027

The recent closing of New York City's Fresh Kills landfill has resulted in the exportation of 13,000 tons per day of New York City Municipal Solid Waste to out-of-state landfills. This paper evaluates integrated waste management options for the city. The first option is source separation of the NYC MSW into two streams: (1) wet, organic food scraps and yard waste and (2) the remainder of the waste stream or the "dry" stream. The wet stream would be processed in a large-scale composting facility to produce a soil conditioner product. The dry stream would be subjected to a highly mechanized Materials Recovery Facility (MRF) which would use best available technology to separate recyclable material such as ferrous and non-ferrous metals, plastics, paper and glass. The remainder of the dry stream would then consist of non-recyclable hydrocarbons which could be used to generate electricity in a Waste-to-Energy (WTE) facility.

Posters

Submission: 33rd Mid-Atlantic Industrial & Hazardous Waste Conference

Presentation Preferred: Poster

Pilot-Scale Study of Fenton's Oxidation of Organic Contaminants in Water

esenting Author: Stephanie Hubbard
Department of Civil and Environmental Engineering
Worcester Polytechnic Institute, Worcester, MA 01609

rresponding Author: John Bergendahl, Ph.D., P.E.
Assistant Professor
Department of Civil and Environmental Engineering
Worcester Polytechnic Institute, Worcester, MA 01609
Email: jberg@wpi.edu; phone: (508) 831-5772

authors: Domenico Grasso, Ph.D., P.E.
Hewlett Professor & Chair, Picker Engineering Program
Smith College, Northampton, MA 01063

Adam Gordon, Worcester Polytechnic Institute

Courtney Estaphan, Worcester Polytechnic Institute

Abstract:

A pilot-scale study was conducted to evaluate an advanced oxidation process ilizing Fenton's reagent for treating contaminated groundwater extracted at a site in onnecticut. Fenton's oxidation uses hydrogen peroxide and iron to produce highly active hydroxyl radicals that mineralize organics in the water. The site for this study as operated for many years as a solvent recovery service (now closed) and resulted in a oundwater contaminant plume containing many volatile organic compounds and cohols. Currently a UV-hydrogen peroxide oxidation system is being operated to treat e extracted groundwater, but high energy and maintenance costs are associated with is treatment system. Through an initial bench-scale treatability study, Fenton's idation was shown to effectively reduce the concentration of contaminants found in is groundwater. The pilot plant is a batch reactor system that is operated to evaluate the fects of various parameters including: pH, iron concentration, hydrogen peroxide dose, d reaction time. Aqueous samples were analyzed by gas chromatography with FID and CD detectors. The data collected have proven the effectiveness of Fenton's oxidation r this contaminated water, with reductions of greater than 90% for some contaminants. ne information gathered from this pilot-scale study will be used to design a full-scale nton's oxidation system.

DEPOSITION OF CHARGED SOOT PARTICLES IN TUBES

ROBERT B. BARAT
Department of Chemical Engineering
New Jersey Institute of Technology
Newark, NJ 07102

JOHN G. STEVENS
Department of Mathematical Sciences
Montclair State University
Upper Montclair, NJ 07043

The new, smaller size (2.5 micron) standard for ambient particulate loadings proposed by the U.S. EPA will require the development of novel particle capture technologies based upon an improved fundamental understanding of fundamental particle transport. In this poster, we consider a model for the deposition of charged particles in an electrically grounded cylindrical tube in the absence of an applied electric field. We consider the underlying particle size distribution, which we discretize into a finite number of fractions. Each fraction has a representative particle size that determines the charge and diffusivity of particles in that fraction. A transport equation is obtained for each fraction by considering the radial electrostatic, radial diffusional, and axial advective fluxes. At short distances into the tube, radial transport toward the wall is dominated by the electrostatic flux of larger particles. At intermediate distances, the contributions of both the diffusional and electrostatic fluxes are important. Farther down the tube, radial transport is dominated by the diffusional flux of smaller particles, with the remaining larger particles persisting in the gas stream. Assuming a single mean particle size, in lieu of a realistic treatment of the actual distribution, results in inaccurate predictions of particle capture. In a realistic laboratory scenario, we flowed a combustion-generated, soot-laden exhaust stream into a tube coated on the inner wall with a platinum oxidation catalyst. An axial high voltage wire provided an applied electric field to enhance soot particle transport to the wall leading to subsequent oxidation. Soot removal was demonstrated by changes in light transmission through filter paper samples. With sufficient oxygen present, higher soot oxidation, as indicated by increased CO_x (CO + CO2) levels and elevated catalytic surface temperatures, was observed in the presence of an applied field. The multi-fraction particle transport model was modified for the applied field, which dwarfs any radial electrostatic flux. Predicted particle penetrations (concentrations) down the tube length were compared well with experimental reactor outlet soot loadings for varying applied voltages.

EPA ENFORCEMENT UPDATE – PREPARE YOUR CAMPUS FOR AN EPA INSPECTION

Adam H. Steinman, Esq.
Vice President of Compliance Services

Woodard & Curran Inc.
41 Hutchins Drive
Portland, ME 04102
800-426-4262
asteinman@woodardcurran.com

ince the late 1990's, the Environmental Protection Agency (EPA) has begun to specifically ocus enforcement efforts against colleges and universities whose activities are regulated by a umber of different environmental laws, and who have not been a primary target for compliance nspections in the past. As a result of the first EPA compliance inspections of colleges and niversities, EPA is now under the impression that there is widespread non-compliance by many f these entities. Consequently, enforcement against these entities, particularly in Regions I, II nd III, has intensified and become, and will continue to be, an Agency priority. Moreover, since 998, the level of fines being levied by EPA against colleges for their non-compliance with nvironmental laws has heightened dramatically.

his presentation will address EPA's enforcement initiatives against colleges and universities in Regions I, II and III; why colleges and universities are vulnerable to enforcement actions; steps olleges and universities can take to reduce the likelihood of an EPA inspection resulting in an nforcement action; strategies for addressing areas of non-compliance; and developing nanagement systems which can be implemented to ensure future compliance.

EXPERIMENTAL DETERMINATION OF ENTHALPIES OF VAPORIZATION OF POLYCHLORINATED BIPHENYLS (PCBs) USING CORRELATION GAS CHROMATOGRAPHY

Swati Puri, William J. Welsh* and James S. Chickos
Department of Chemistry and Biochemistry, University of Missouri–St. Louis
8001 Natural Bridge Road, St. Louis, Missouri 63121
*Corresponding Author: Email Address: wwelsh@umsl.edu

Polychlorinated biphenyls (PCBs) are persistent organic contaminants that are found at an appreciable concentration in the polar regions, presumably as a result of long-range atmospheric transport. For many environmental applications, vapor pressure and enthalpy of vaporization are the relevant thermodynamic properties associated with the dispersal of PCBs.

The vaporization enthalpies of 16 PCBs have been determined by correlation gas chromatography. This study was prompted by the realization that the vaporization enthalpies previously reported by other workers (Falconer, R.L.; Bidleman, T.F. *Atm. Env.* **1994**, *28*, 547-554) for the standard compounds, octadecane and eicosane, were measured at 340 and 362 K, respectively, rather than at 298 K. An adjustment to 298 K results in a 7-8 kJ/mol increment in the vaporization enthalpies of PCBs. With inclusion of this adjustment, vaporization enthalpies evaluated by correlation gas chromatography are in good agreement with the values determined previously in the literature. The vaporization enthalpies of PCBs increased with the number of chlorine atoms and were found to be larger for the meta- and para- substituted PCB isomers.

Correlation gas chromatography is a fast and reliable method of evaluating enthalpies of vaporization because it is based on the vaporization enthalpies of several standard compounds, such as *n*-alkanes and various chlorinated hydrocarbons, whose values are well established in the literature.

This research is partially funded by a grant to WJW from the U.S. Environmental Protection Agency's Science to Achieve Success (STAR) program.

Managing Municipal Solid Wastes: Japan, U.S. and EU

Ko Matsunaga and Nickolas J. Themelis

Earth and Environmental Engineering Department
Henry Krumb School of Mines
Columbia University in the City of New York
New York, NY 10027

e management of Municipal Solid Waste (MSW) is becoming a top priority task for all governments. t the amount of MSW varies from countries. Economic power has affected it, but through research, found there are other factors except GNP. Our research has found new parameters " pressure of MSW" th GNP and population density, which determines the management of MSW. The "high pressure" untries are required to solve the MSW due to land limitation and sanitation. Proof of "Pressure of MSW" that South Korea is establishing better MSW management than the U.S. Other expected "hot spots" iwan, Puerto Rico, and other islands.

Understanding of PCB Structural Properties Using Classical Thermodynamics and Computational Modeling

Robert Ambrosini[1], David J. Braun[1],
Nada Assaf-Anid[1], Lisa Totten[2]

Manhattan College[1]
Rutgers University[2]

ABSTRACT

In recent years, the role of chlorinated organic pollutants in the environment has received increasing attention in both the scientific community and among the general public. This interest has inspired a considerable number of on-going studies directed towards the reductive dechlorination of aliphatic and aromatic pollutants such as carbon tetrachloride and polychlorinated biphenyls (PCBs). In spite of these efforts, the complete thermodynamic data for compounds such as PCBs is largely unavailable. It is through the collection of this data that we can enhance our understanding of the thermodynamics and kinetic feasibility of these reductions. In this work, computational chemistry calculations using the TITAN computer program employing the B3LYP/6-31G* method are used for the purpose of potentially filling these gaps in our information. Derived parameters include dihedral twist angles, free energies, and lowest unoccupied molecular orbital (LUMO) energies. These parameters are then related to the published results of classical thermodynamic calculations with the eventual objective of correlating thermodynamic descriptors to rate constants for reactions of PCBs with various reductants.

Non-Point Source Nutrient Loads from Lower Westchester County, NY, into Long Island Sound

Nada Assaf-Anid*[1], XueFeng Liu[2], Joshua Hutson, Scott Lowe[3] and John Jeris[3]

[1]Chemical Engineering Department
Manhattan College, NY
Riverdale, NY 10471
Email: nassafan@manhattan.edu

[2]Earth and Environmental Engineering Department
918 Seeley Mudd Building, Columbia University
500 West 120th St, New York, NY 10027
Email: xl73@columbia.edu

[3]Environmental Engineering Department
Manhattan College, NY
Riverdale, NY 10471

.BSTRACT

utrophication has been identified as the most pressing water quality problem in the Long Island ound due to excessive nutrient loading from land-based point and non-point sources and from tmospheric deposition. A three-year Water Quality Monitoring Program was initiated in anuary 1997 by the Westchester County Department of Planning and New York State)epartment of Environmental Conservation (NYSDEC) to determine the nutrient loads delivered) Long Island Sound from the two largest watersheds in lower Westchester County via Aamaroneck and Blind Brook rivers. Baseline and storm events were studied in an attempt to btain representative samples that included both dry and wet weather conditions. Water emperature, pH, conductivity, dissolved oxygen and flow rate were monitored in the field, hereas turbidity, total suspended solids, total coliform, fecal coliform, total nitrogen, total Kjeldahl nitrogen (TKN), nitrate, nitrite, ammonia, total phosphorus, and orthophosphate were nalyzed. Annual input of nutrients from the Mamaroneck and Blind Brook watersheds into the ong Island Sound were determined using water concentrations data and river flows. Although ne Mamaroneck and Blind Brook Wastewater Treatment Plants are still the main contributors of itrogen loading, 217 tons and 29 tons from both MWWTP every year respectively, the data how that non-point sources from both watersheds contribute significant loads amounting to 70 ons and 40 tons, respectively. In Spring 2000, rainfall samples collected in two locations in the watersheds during three storm events show that total nitrogen loads from atmospheric sources ould account for 14% and 18% of the total loads measured in Mamaroneck River and in Blind Brook respectively.

MATHEMATICAL MODELING OF TOXICS FATE, TRANSPORT AND BIOACCUMULATION USING GIS

FERDI HELLWEGER
HydroQual, Inc.
1 Lethbridge Plaza
Mahwah, NJ 07430

ABSTRACT

A toxic chemical fate, transport and bioaccumulation system has bee developed by linking EPA's WASTOX FORTRAN code ArcView. The systen called GISTOX, is fully compatible with EPA's BASINS database. River reac information, water quality data and point source loads are automaticall extracted from the database. The system supports solids sorption, volatilizatior biodegradation and hydrolysis fate processes. The present version of the syster is restricted to steady state transport in one dimension. The bioaccumulatio part of the model is set up using a preset food web structure consisting of benthi invertebrate, phytoplankton, zooplankton, small fish and large fish.

THE EXAMINATION OF SODIUM AND CHLORINE AS FLAME INHIBITORS IN A METHANE/AIR FLAME

ANN MARIE FLYNN
Department of Chemical Engineering,
Manhattan College,
Riverdale, NY, 10471

INTRODUCTION

The Montreal Protocol effectively banned the use of many halogenated products such as Freons due to their adverse effect on the ozone layer. Related halo-compounds serving as flame-retardants are also being replaced. While effective as retardants, these compounds often produce toxic byproducts during use as fire-fighting agents (1). At present, considerable effort is being spent to find alternative fire retardants.

Earlier combustion literature indicates that sodium compounds were once considered for flame retardation. Iya and coworkers showed that sodium inhibition, as measured by temperature rises in post flame gases of quenched CH_4/air flat flames and by OH absorption spectroscopy, was due to a homogeneous gas phase chemical mechanism in which peak OH radical concentrations are reduced (2). They also suggested that Na atoms catalyze radical recombination. Vanpee and Shirodkar showed that organo-sodium compounds were effective flame inhibitors, as measured by an increased oxygen mass fraction at extinction in an opposed jet liquid fuel burner (3). More recently, Glazkova et al. observed that Na salts inhibited CH_4/air deflagrations (4). Kim found that certain Na salts are effective suppressants of high temperature flames (5). In addition, sodium compounds form a class of flame retardants which are ozone-friendly.

Indeed, the alkali metal and salts have been considered an important and not fully explored class of flame inhibitors (6). The purpose of this work was to investigate the effects of sodium when compared to chlorine as a chemical inhibitor on a methane/ air flame. It is possible that sodium could be a replacement for the common, commercially available flame retardants that are based on the halogenated organic compounds (HALONS)—neat or in combination with phosphorous containing organics or antimony oxide (3).

The following sections will examine how sodium compares to chlorine as a chemical inhibitor based on temperature profile crossovers, reduction of CO burnout (as CO/CO_2 ratio), laminar burning velocities, inhibition index, and sodium concentration profiles.

EXPERIMENTAL SECTION

Figure 1 presents a schematic diagram of the experimental apparatus used for optical data collection. The burner was a stainless steel, slotted type, typically found in an atomic absorption spectroscopy (AAS) apparatus, which had a slot 0.07 cm wide and 10 cm long. It was chosen because it provided a long path for light absorption measurements. The burner was press fitted into a customized PVC mixing chamber and the chamber was then mounted onto an X-Y translation stage to provide two-dimensional spatial resolution of the flame relative to the stationary light source. The mixing chamber was fitted with two ports on either side of the mixing chamber, offset from each other, to accommodate the introduction of the primary (methane) and an auxiliary (methyl chloride) fuel or oxidant (air or oxygen). The offset allowed the components to combine in a swirling fashion and encouraged more complete mixing. An air atomizing nozzle (Hago) attached to the mixing chamber was used to supply a 0.02 M sodium nitrate ($NaNO_3$) solution to the flame. The light source was a spectrally broad, continuum light source (Oriel 6259 300 W xenon arc lamp) because future plans for this apparatus included the detection of atomic species other than sodium as well as diatomics. The light source was apertured by a water filter followed by a 5 mm. aperture before it passed through the flame so as to increase spatial resolution within the flame (see Figure 1). A Spex 1400-11 Double Spectrometer with a side-on EMI/Thorn Photomultiplier Tube (PMT) was used for spectral resolution.

A Linear chart recorder was used to record the signal from the PMT. The optical measurements were converted to absorbance measurements in the following manner.

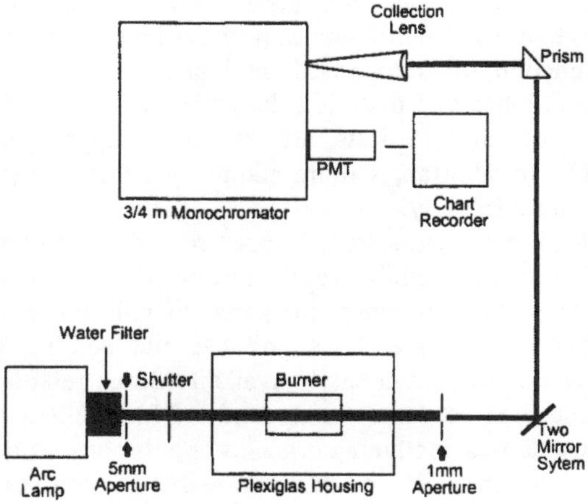

FIGURE 1. Schematic Diagram for Optical Data Collection

A dark signal, D, a signal due to the incident source radiation, V_0, and a signal due to attenuation of light by the sample, V, were used to calculate absorbance:

$$A = \ln\frac{I_0 - D}{I - D} \qquad (1)$$

Flame temperatures were measured with an un-coated, Type R thermocouple (Pt, Pt/13% Rd), having a bead diameter of approximately 0.5mm. The flame temperatures were then corrected to account for radiation heat loss to the surroundings and conduction heat loss along/through the thermocouple wires.

In order to determine stable species concentrations, such as CO, CO_2, and CH_4, hot gases were drawn from the flame with a water-cooled, quartz microprobe (50 μm orifice) through a heated transfer line into a Perkin-Elmer 3920 gas chromatograph (GC) fitted with a flame ionization detector (FID). A schematic diagram of the apparatus used to collect the stable species data is shown in Figure 2. A Hewlett-Packard 3396 Series II Integrator was used to process the FID signal and record the results.

The entire burner and mixing chamber assembly, quartz probe for gas sampling, and thermocouple were housed in a vented, five sided, transparent Plexiglas structure (18"x18"x24").

RESULTS AND DISCUSSION

A variety of indicators quantify the inhibition effectiveness of a particular species on a flame. A chemical inhibitor is characterized by the small amounts of an agent that are needed to retard or extinguish a flame (5). By contrast, a physical inhibitor is characterized by the large quantities of agent required to extinguish a flame, e.g., water.

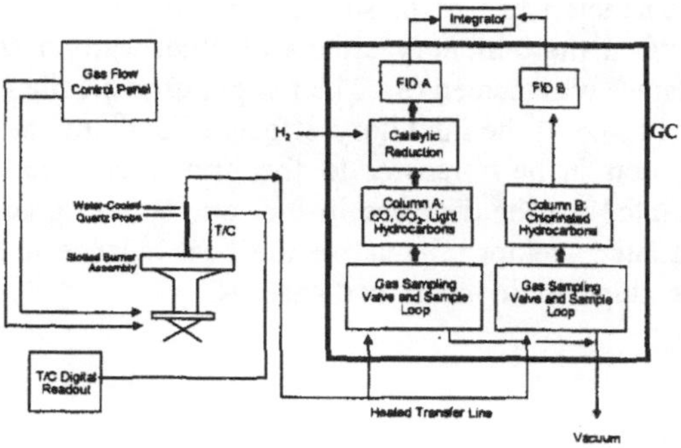

FIGURE 2. Schematic Diagram for Stable Species Data Collection

It has been shown that sodium salts act as effective chemical inhibitors when introduced to various hydrocarbon flames (2, 3, 5, 6, 7). In addition, chlorinated hydrocarbons have been studied because of their inhibition effects on hydrocarbon flames (8, 9, 14). The inhibition effectiveness of chlorine or sodium addition to CH_4/air flames has been expressed in many ways. The following measurement tools were used to characterize and/or compare the effectiveness of these species as chemical inhibitors.

Temperature Profile Crossover

It has been shown that chlorinated, laminar, flat flame temperatures are slightly lower in the preheat zone and can achieve a higher maximum temperature at a greater height above burner (HAB) relative to hydrocarbon flames (9). This particular phenomenon of chlorinated flames exhibiting slightly higher post flame temperatures as compared to methane flames has been explained by Kaskan et al. and by Kim (2, 5, 6). In essence, when a quenched flame burning at constant velocity is inhibited, some one or more critical reactions are impeded. This causes the flame to burn less closely to the burner, to lift off slightly, thus reducing the temperature gradient and heat transfer at the burner surface. As a result the gas temperatures throughout the flame are increased and so, presumably are, the rates of the critical reactions, previously slowed by the inhibitor, increase to the point where they can continue to consume the constant throughput of fuel and air. The net reduction in heat transfer then results in a temperature rise in the unburned gases. Therefore a measurement of inhibition of a quenched premixed flame burning at a constant, unburned gas velocity is a rise in temperature in the inhibited flame (e.g., CH_4/CH_3Cl/Air), relative to the uninhibited flame (e.g., CH_4/Air). This is signaled by a crossover in the temperature profiles of the inhibited versus the uninhibited flames, which was used as an indication of chemical inhibition behavior.

Figures 3-5 shows that a crossover in the temperature profile exists for a methane/air flame that has been doped with sodium, chlorine, and sodium and chlorine together, respectively.

The crossover characteristics of these temperature profiles mimic the results of Barat et al. when the inhibitory effects of chlorocarbon addition to fuel-rich methane/air flames was studied (8). The temperature profile crossovers in this work provide evidence of the inhibitory effects of sodium and chlorine. As expected, the reduction in heat transfer to the burner as a result of the inhibitor addition accounted for the decrease in the temperature gradient early on in the profile and a later attempt to equalize the heat balance in the flame accounted for the temperature rise in the burned gases (6).

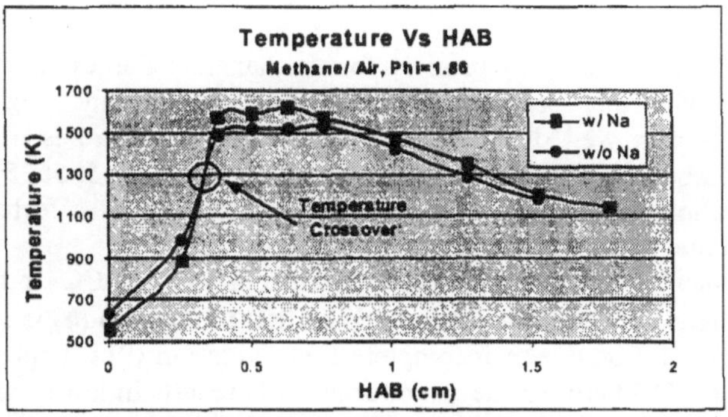

FIGURE 3. Illustration of Temperature Profile Crossover for CH_4/Air Flame Doped with Sodium Inhibitor, ϕ=1.86

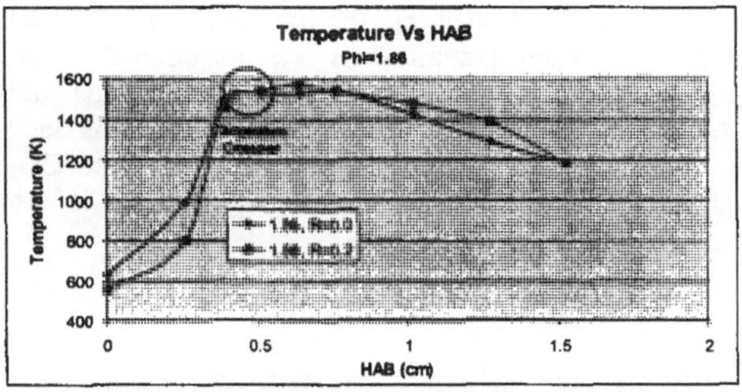

FIGURE 4. Illustration of Temperature Profile Crossover for a CH_4/Air Flame Doped with Chlorine Inhibitor (fed as CH_4/Air/CH_3Cl), ϕ=1.86

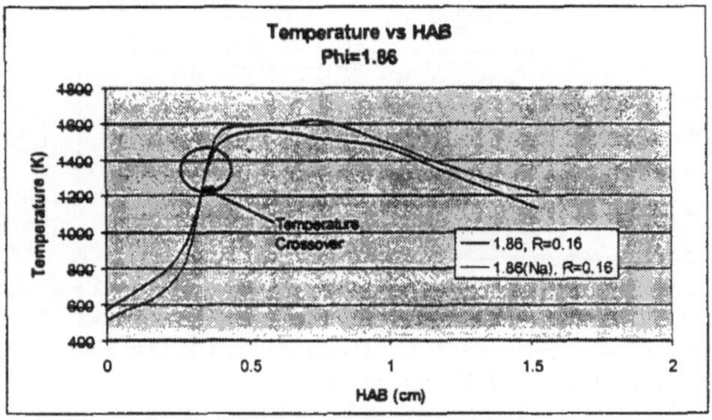

FIGURE 5. Illustration of Temperature Profile Crossover for CH_4/Air Flame Doped with Chlorine Inhibitor When Compared to CH_4/Air Flame Doped with Chlorine and Sodium Inhibitors, ϕ=1.86

CO/CO_2 Profiles

Figures 6-8 depict CO/CO_2 profiles for methane/air flames that have been doped with chlorine, sodium, and sodium and chlorine together, respectively. It has been well documented that the CO/CO_2 ratio is a good indicator of hydrocarbon combustion inhibition or efficiency (51, 55, 56, 57). In addition, concentration ratios, such as CO/CO_2, should be less affected by changes in dilution due to air entrainment into the flame.

The experimental data showed that the rise in the CO/CO_2 ratio (a function of the decrease in CO burnout) for a fixed fuel to air ratio (ϕ) is strong evidence of flame inhibition due to incomplete fuel oxidation (9).It is possible that a competition for OH between the inhibitor and CO results in lower rates of CO conversion to CO_2.

The experimental data depicted in Figure 6 shows that the CO/CO_2 ratio increased when chlorine-alone (R=0.2) was added to a methane/air flame up to an HAB of approximately 0.8 cm. After this point, the CO/CO_2 ratio is lower in the chlorinated flame. Figure 7 indicates that the increase in the CO/CO_2 ratio was much more significant when sodium-alone was introduced to the same flames. Indeed, the CO/CO_2 ratio is consistently higher in the flame containing sodium. In contrast, Figure 8 shows that when chlorine (R=0.2) is added to a methane/air flame that has been doped with sodium, the CO/CO_2 ratio *decreases*. This was an unexpected result and its significance is discussed below.

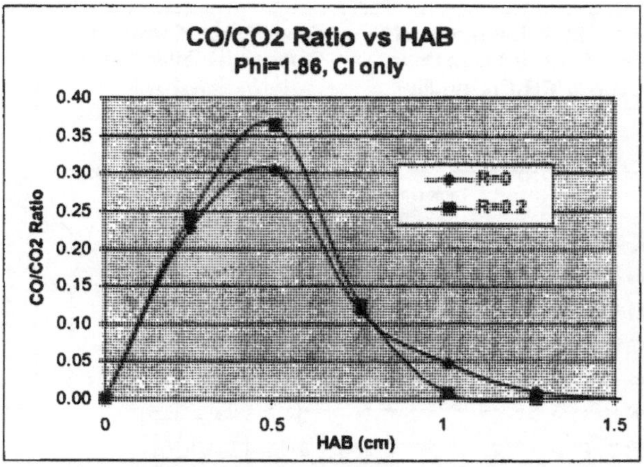

FIGURE 6. Experimental CO/CO_2 Ratios for CH_4/Air Flame Doped with Sodium Inhibitor, ϕ=1.86

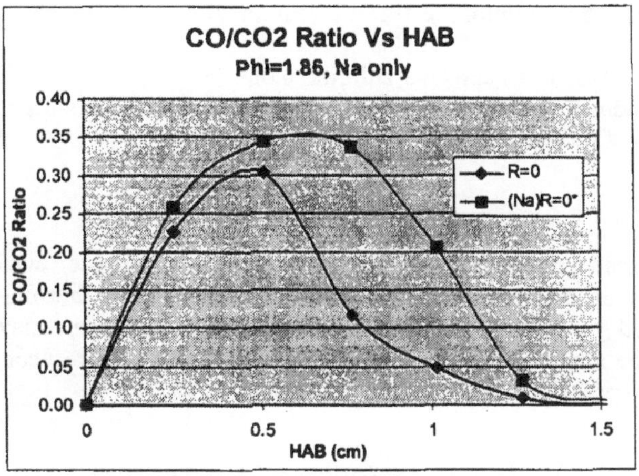

FIGURE 7. Experimental CO/CO_2 Ratios for CH_4/Air Flame Doped with Chlorine Inhibitor, ϕ=1.86

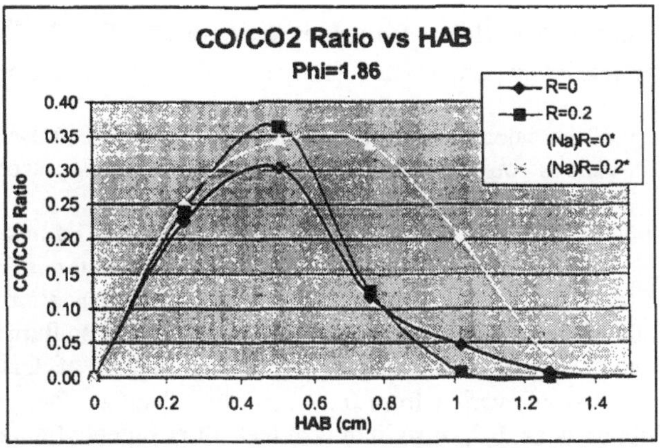

FIGURE 8. Experimental CO/CO_2 Ratios for CH_4/Air Flame Doped with Chlorine, Sodium with Chlorine, and Sodium Inhibitor(s), ϕ=1.86

Both sodium and chlorine have inhibitory effects on a methane/air flame when introduced to the flame separately. Indeed, the results show that any combination of sodium and/or chlorine causes the CO/CO_2 ratio to rise, with respect to the uninhibited flame. It should be noted that based on CO/CO_2 ratios as inhibition indicators, the inhibitory effect of sodium-alone is more significant than the inhibitory effect of chlorine-alone. However, as regards CO burnout (i.e., conversion of CO to CO_2), it appears that sodium and chlorine (R=0.2) together are less effective as an inhibitor than sodium alone. This implies that chlorine decreases the inhibitory effect of sodium alone, due to Na and Cl interactions.

Therefore, it is concluded that chlorine alone (R=0.2) is a less effective inhibitor than sodium alone and that chlorine serves to reduce/impede the inhibitory effects of sodium.

Inhibition Index

Kim defines an inhibition index, I^T_Q, as a numerical index for evaluating the effectiveness of inhibitors in quenched, premixed flames burning at a constant unburned gas velocity (5). The index compares a rise in the maximum temperature in the inhibited flame to the uninhibited flame with respect to the inhibitor on a per mole of inhibitor basis. It is defined as:

$$I_Q^T = \frac{T_i^{Max} - T_0^{Max}}{[I]} \qquad (2)$$

where T_i^{Max} is the inhibited flame maximum temperature [K]; T_0^{Max} is the uninhibited flame maximum temperature [K]; and [I] is the inhibitor mole fraction.

A summary of the calculated inhibition indices for sodium and chlorine when combined in various configurations in methane/air flames is presented in table 1.

When comparing a flame doped with chlorine alone and flames doped with sodium alone, the ΔT value of the numerator of the inhibitor index i.e., $T_i^{Max} - T_o^{Max}$, at least triples. The constant molar flow rate of sodium to all flames was 4.2E-06 moles/min. As a result, the mole fraction of sodium varied from 2.18 E-05 to 2.19 E-05. In contrast, the mole fraction of CH_3Cl in methane/air flames (ϕ=1.86) varied from 0.87 E-02 to 3.5 E-02. These results show that the inhibitor index, I_Q^T, of sodium is at least three orders of magnitude *larger* than the inhibitor index of chlorine. The ΔTs for the sodium flames were much larger than those for the chlorine flames. This alone would contribute to the inhibitor index being larger for sodium than for chlorine. However, the largest contribution to the difference in inhibition indices was the fact that the mole fractions for sodium were three orders of magnitude *smaller* than the mole fractions of chlorine. In effect, these results imply that a very small amount of sodium (supplied as Na^+) would produce the same (if not larger) inhibitory effect on a methane/air flame as a relatively large quantity of chlorine (supplied as CH_3Cl).

TABLE I. Summary of Inhibition Indices of Sodium and Chlorine in CH_4/Air/Na, CH_4/Air/CH_3Cl, and CH_4/Air/CH_3Cl/Na Flames, ϕ=1.86

Inhibitor Present	Max. Temperature (K)	$T_i^{Max} - T_0^{Max}$ (K)	I_Q^T [K/ mole fraction]
None (uninhibited)	1531	0	0
Na	1616	85	3.87E06
Cl (R=0.07)	1547	16	1.44E03
Cl (R=0.11)	1543	12	6.74E02
Cl (R=0.16)	1561	30	1.16E03
Cl (R=0.2)	1564	33	1.00E03
Na + Cl (R=0.16)	1616	85	3.29E03 (due to Cl) 3.89E06 (due to Na)
Na + Cl (R=0.2)	1630	99	3.01E03 (due to Cl) 4.54E06 (due to Na)

Sodium Concentration Profiles

Sodium absorbance profiles were measured for flames that were doped with sodium and sodium plus chlorine at varying ratios. These absorbance profiles were then converted to concentration profiles. Figures 9-10 presents the Na profiles for varying equivalence ratios.

Figure 9 shows that the sodium atom profiles are typical of the behavior of alkali and alkaline earth metals in a flame i.e., the atom concentration quickly climbs to a maximum and then drops off as oxides are readily formed (13). The shift of maximum sodium atom production to higher HAB as the equivalence ratio increases and the flame speed decreases is due to the maximum temperature shifting to higher HAB as the equivalence ratio increases.

Figure 10 presents Na absorbance versus HAB profiles as the chlorine to sodium ratio is increased.

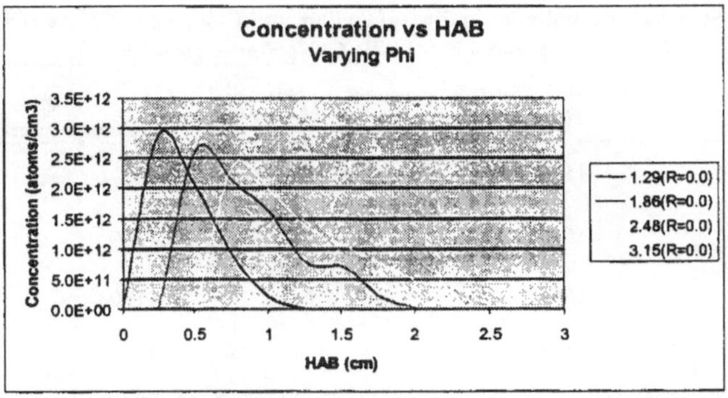

FIGURE 9. Sodium Concentration versus HAB for Methane/ Air/Sodium Flames as Equivalence Ratio is Varied

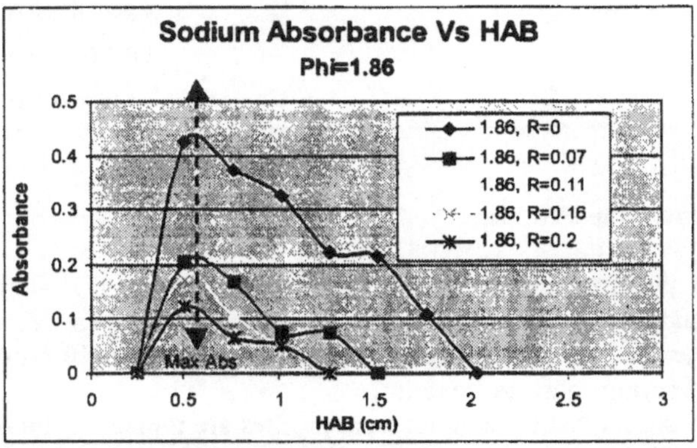

FIGURE 10. Experimental Sodium Absorbance Profiles for CH_4/Air/CH_3Cl/Na Flames, ϕ=1.86, as the CH_3Cl Ratio is Varied

A very interesting conclusion can be drawn from the data. It would have been expected that the addition of a chemical inhibitor such as chlorine to a methane/air flame that had been doped with sodium would result in the impedance of some of the chemical reactions and an ultimate delay in the maximum production of sodium atoms i.e., higher HAB. Instead, it is clear that that the addition of chlorine to a methane/air flame doped with chlorine serves to drastically reduce the net production of sodium atoms in the flame. For example, when the fuel contained as little as 7% chlorine, the maximum sodium atom production was reduced by approximately 50%. When the fuel contained 20% chlorine, the sodium atom production was reduced by 75%. Clearly, instead of the typical independent pathways of Na→NaO, NaO_2, NaOH and Cl→HCl, it is concluded that the more prevalent combined pathway was Na+Cl→NaCl.

Literature Cited

1. McNesby, K.L., R.G. Daniel, A.W. Miziolek, and S.H. Modiano, *Applied Spectroscopy,* vol. 51, p. 678-683 (1997)

2. Iya, K., Wollowitz, S. and Kashan, W., *Combust. Flame*, 22, 415-417 (1974)

3. Vanpee, M. and Sherodkar, P., *Seventeenth Symposium (Int) on Combustion*, 787-795 (1978)

4. Glazkova, A.P., V.P. Karpov, and P.V., Phil, *Archivum Combustionis,* vol. 11, p. 273-282, (1991)

5. Kim, H.T., *Korean J. of Chem. Eng.*, 9(1), 1-7 (1992)

6. Iya, K., Wollowitz, S. and Kashan, W., *Symp (International) Combust.*, [Proc], 15, 329-336 (1975)

7. Jensen, D.E., G.A. Jones, *J. Chem. Soc., Faraday Trans. I,* 78, p. 2843-2850, (1982)

8. Wang, L. and Barat, R.B., *Hazardous Waste and Hazardous Substances*, vol. 12, p51+ (1995)

9. Wang, L., Jalvy, P., and Barat, R.B., *Combust. Sci. and Tech.*, vol. 97, p. 13-36 (1994)

10. Barat, R.B., A.F. Sarofim, J.P. Longwell, and J.W. Bozzelli, *Combustion Science and Technology,* vol. 74, p.361 (1990)

11. Brouwer, J., J.P. Longwell, A.F. Sarofim, R. B. Barat, and J.W. Bozzelli, *Combustion Science and Technology,* vol. 85, no. 1-6, p. 87, (1992)

12. Ho, W.P., Q.R. Yu, and J.W. Bozzelli, *Combustion and Flame,* vol. 88, p. 265, (1992a)

13. Skoog, D.A., Principles of Instrumental Analysis, Saunders College Publishing (1985)

14. Valeiras, H., Gupta, A.K. and Senkan, S. M., *Combust. Sci. and Tech.*, vol. 36, p.123-133 (1984)

ANAEROBIC TRANSFORMATION AND BIOREMEDIATION OF CEPHALEXIN ANTI-BIOTIC WASTE SLUDGE USING A FLUIDIZED BED REACTOR

R. SARAVANANE

D. V. S. MURTHY

K. KRISHNAIAH

Environmental Engineering Laboratory,
Department of Chemical Engineering,
Indian Institute of Technology, Madras,
Chennai - 600 036
India

INTRODUCTION

Cephalexins are widely manufactured and used for the therapy of acute pneumonia and infections of respiratory tract, skin and soft tissues, bones and joints, urinary tract and blood stream [1]. Cephalexin ($C_{16}H_{17}N_3O_4S.H_2O$) is an intermediate of cephalosporin [2]. Since the biotransformation of cephadine and cephalexin gives the same acylated products, they are being referred by a single name as cephalexin throughout the manuscript. Pharmaceutical industries manufacturing cephalexins based drug, suffer from inadequate treatment and disposal due to high concentration of cephalexin (120 to 175 mg/L) and complex biomass resulting from their reaction with organic solvents and volatile solids available in the combined effluent of the plant. The presence of cephalexin increases total strength (Chemical Oxygen Demand) of the effluent and indirectly increases the cost of the treatment. Hence the biotransformation of cephalexin either alone or in combination with other energetic compounds, offers the potential for an economical and environmentally friendly disposal alternative.

Microbial biotransformation of antibiotics has been reported [3] for the preparation of analogs of members of several classes of antibiotics. Of these transformation reactions, acylation [4] of 6-aminopencillanic acid (6-APA), 7-aminocephalosporanic acid (7-ACA) and 7-amino-3-deacetoxycephalosporanic acid (7-ADCA) have been of prime importance in the biotransformation of cephalexins. The cleavage of β-lactum ring structure facilitates further degradation resulting in the formation of a linear structure and loss of antibiotic activity. Ring cleavage of cephaloridine and cephalothin have been reported by researchers [5-7] for the ring cleavage of β-lactum. Bioaugmentation refers to a method of maintaining sufficient biomass when adequate carbon substrates and nutrients are unavailable in the pharmaceutical effluent. Bioaugmentation using purchased cells [8] could not find evidence of successful operation with respect to anaerobic degradation of industrial effluents. The present research follows the research work [9] using suspended cultures, where an off-line reactor, called an enricher-reactor [10] used for anaerobic degradation of different constituents of antibiotic and synthetic drug based effluents under methanogenicand iron reducing conditions. The objectives of the study were to evaluate the effect of bioaugmentation on cephalexin removal using anaerobic cephalexin degrading enrichment culture as inoculum and to examine the effect of hydraulic retention time (HRTs) and cephalexin concentration on the extent of cephalexin transformation.

MATERIALS AND METHODS

Enrichment Procedures for Cephalexin Degrading Reactor

Semicontinuous cephalexin degrading reactor was maintained in a stationary position at 28 °C in 700 mL serum bottles containing 500 ml of liquid. The serum bottles were used to acclimatize the enrichment culture (procured from Southern Pharmaceuticals Pvt. Ltd., Tamilnadu, India) with the pharmaceutical effluent to be treated. After 6 days 20 % of well-mixed liquid was removed and replaced with fresh feed. At this time the bottles were purged with N_2 gas in order to maintain anaerobic conditions. The fill and draw operation was conducted every 6 days. The same company supplied the pharmaceutical effluent and the characteristics were presented in Table I. The influent to the fluidized bed reactor is the pharmaceutical effluent. During the concentration study, deionized water was used to dilute the wastewater concentrations of cephalexins.

In Table I, the parameters measured as per Standard Methods [11] include BOD_5, COD, Total suspended solids, Total dissolved solids, Total volatile solids, ammonia and surfactants. The analysis of cephalexin, 7-amino-3-deacetoxycephalosporanic acid (7-ADCA) and acyl group (Phenyl acetic acid) was carried out as described in British Pharmacopoeia [2] using an Infrared Spectrometer (Model M - 500). The acyl group refers to Phenyl acetic acid. In the present study, the biotransformatin of cephalexin results in the products viz. 7-amino-3-deacetoxycephalosporanic acid (7-ADCA) and acyl group (Phenyl acetic acid). Hence the measurement of acyl group is related to the influent concentration of cephalexin. Pure cephalexin and 7-amino-3-deacetoxycephalosporanic acid (7-ADCA) conforming to British Pharmacopoeia Chemical Reference Substance, (BPCRS) [2] were used for preparing their calibration graph. The methane content of biogas was analyzed by a gas chromatography (GC) on a Hewlett-Packard, model 5890 A, equipped with a flame ionization detector (FID).

Bioaugmentation was carried out using an enricher-reactor where optimum growth conditions were maintained. The concentrations of volatile fatty acids and sucrose in the enricher-reactor were 200 mg/L and 52 mg/L respectively. They were constant throughout the study. The enricher-reactor also received the nutrient solution composed of: KH_2PO_4 (720 mg/L), NH_4Cl (180 mg/L), $MgSO_4$ (152 mg/L), $CaCl_2.2H_2O$ (15 mg/L). Activated carbon particle of average size 700 μm and density 1500 kg/m^3 was used as carrier material.

The reactor column was 7.5 cm in diameter and 120 cm in height (Figure. 1). A gasholder of diameter 15 cm and height 30 cm was mounted on the top of the reactor column. The spatial liquid phase volume of the reactor was 11 L. The feed pumping system consists of two mono block pumps (each 0.25 hp) operated by means of a digital timer system. The reactor was fluidized at bed expansions varying from 30% to 40%. The minimum fluidization velocity for the carrier (activated carbon particle of average size: 700 μm and density 1500 kg/m^3) was found to be 13.2 m/hr. The vertical flow velocity (fluidization velocity) was varied from 13.2 to 20 m/hr.

TABLE I. CHARACTERISTICS OF COMBINED STREAM PHARMACEUTICAL EFFLUENT

Sl.No.	Characteristics	Value
1.	Flow rate (m^3/d)	350
2.	pH	3-4
3.	Temperature (°C)	25 - 45
4.	COD (mg/L)	12000-15000
5.	BOD_5 (mg/L)	2000
6.	Cephradine (mg/L)	200-315
7.	Total Suspended solids(mg/L)	6000
8.	Total Dissolved solids(mg/L)	11000-18500
9.	Total Volatile solids(mg/L)	7500
10.	NH_3 (mg/L)	15 - 40
11.	Surfactants	2 – 20

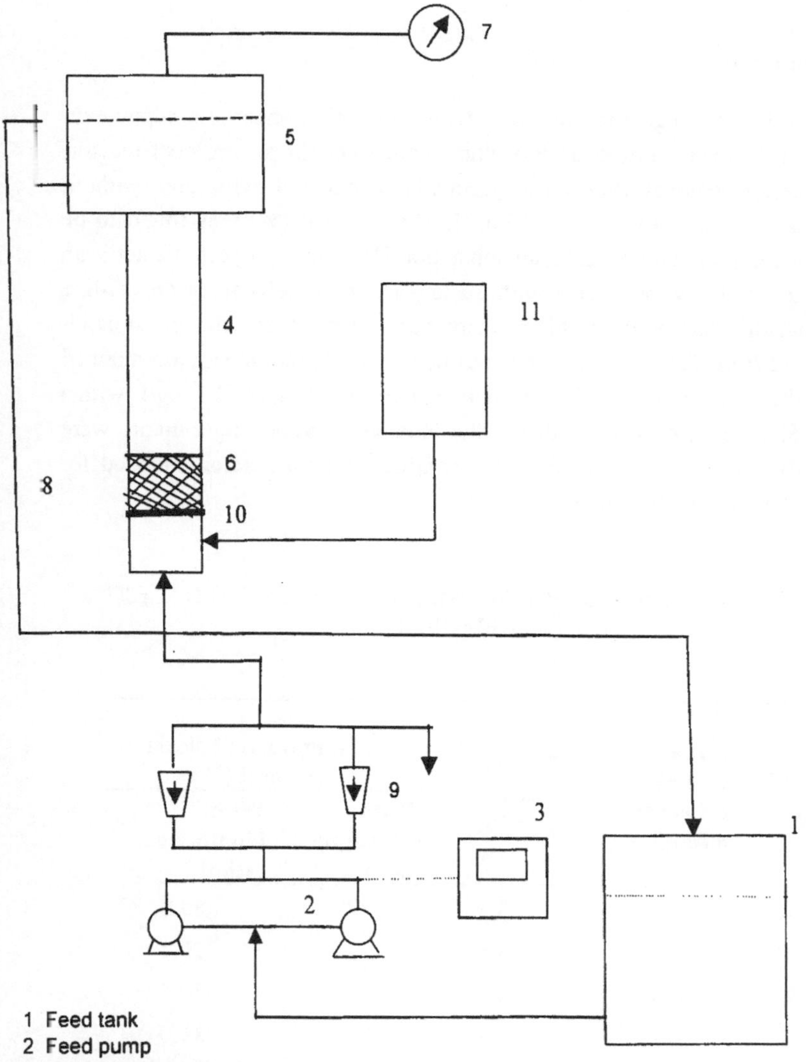

1 Feed tank
2 Feed pump
3 Digital timer system
4 Reactor column
5 Gasholder
6 Carbon bed
7 Wet gas meter
8 Recycle line
9 Rotameter
10 Wire mesh
11 Enricher reactor

Figure 1. Schematic Diagram of Anaerobic Fluidized Bed Reactor

RESULTS AND DISCUSSION

Bioaugmentation

The method of bioaugmentation was carried out using anaerobic cephalexin degrading enrichment culture as inoculum maintained in an enricher-reactor. The cephalexin removal (%) for experiments conducted with and without bioaugmentation was presented in Table II. The removal (%) was found to be higher for the experiments with bioaugmentation [10] (through periodic addition of acclimated cells every 2 days with 30 to 73.2 g of Cells from an off-line enricher-reactor) than without bioaugmentation (without enricher-reactor). It was observed from Table II that the experiments with bioaugmentation attained a removal (%) of 81 at a maximum cephalexin concentration of 175 mg/L with a minimum HRT of 3 h. The results of the bioaugmentation experiments were subsequently used for the start-up of the continuous flow reactor intended for cephalexin transformation studies.

TABLE II. THE EFFECT OF BIOAUGMENTATION ON CEPHALEXIN REMOVAL (%)

Expt	**HRT (h)**	**Initial cephalexin concentration (mg/L)**	**Biomass added[a] (g VSS)**	**Maximum cephalexin removal (%)**	
				Without bioaugmentation	**With bioaugmentation**
E1	36	10	30.2	60	85.2
E2	15	37.8	43.5	54.6	79.2
E3	5.3	137.2	64.7	46.2	80.4
E4	3	175	71.3	38.5	81.0

[a]For bioaugmentation experiment only.

Biotransformation of cephalexin by anaerobic fluidized bed reactor

EFFECT OF HRT ON CEPHALEXIN TRANSFORMATION.

The experimental results of the effect of HRT on cephalexin transformation and its acylated products were presented in Table III. The cephalexin transformation and its by-products are shown below:

$$\text{Cephalexin} \xrightarrow[\text{Transformation}]{\text{Anaerobic}} \underset{\text{(7-amino-3-deacetoxy cephalosporanic acid)}}{\text{7-ADCA}} + \underset{\text{(Phenyl acetic acid)}}{\text{Acyl group}}$$

Low concentrations of cephalexin (2.8 to 3.5 mg/L) were chosen to study the possibility of complete transformation. The analysis was conducted normally every two days till run R6, while almost every day in run R7. First the presence of 7-amino-3-deacetoxycephalosporanic acid (7-ADCA) was confirmed at the out let. Then the continuous feeding of the cephalexin was begun. The first influent concentration of 2.7 mg/L was fed at a HRT of 76 h operated for a period of 6-36 days. Then the HRT was gradually reduced to 5.3 h with cephalexin concentration slightly increased to a maximum of 3.5 mg/L. The removal efficiency of cephalexin was found to be almost same at all HRT above 15 h and attained 82.2% after 102 days of operation. It was inferred from Table 3 that 7-ADCA was the major transformation product formed as compared to acyl group (phenyl acetic acid) with respect to influent cephalexin concentrations.

EFFECT OF CEPHALEXIN CONCENTRATION ON REACTOR PERFORMANCE.

The effect of higher influent cephalexin concentration on reactor performance was assessed in terms of cephalexin transformation and its removal efficiency (%) with influent and effluent data for the entire experimental period were plotted in Figure.1.The influent cephalexin concentration was increased from 5.6 to 175 mg/L with HRT decreased from 76 to 3 h. The 7-amino-3-deacetoxycephalosporanic acid (7-ADCA) was the major transformation product in these runs. The removal efficiency (%) of cephalexin corresponding to maximum concentration of 175 mg/L was found to be 81. The performance of the reactor was also assessed by the volatile acid and methane yield as represented in Figure 2. The methane yield was found to increase with increase in concentration and days of operation and reaches a maximum at cephalexin concentration of 220 mg/L. Volatile acid was found to follow the profile of acyl group and becomes steady at the end of 90 days. It was evident from Table III that the effectiveness of treatment (removal efficiency (%)) was the same at all HRTs above 15 h and remained to be constant with increasing influent concentration. Hence this would be of practical significance in choosing the efficient and economical operating point for the treatment of cephalexin based pharmaceutical wastewater.

TABLE III. EXPERIMENTAL RESULTS OF CEPHALEXIN TRANSFORMATION UNDER EACH HRT CONDITION

Run		R1	R2	R3	R4	R5	R6	R7
Days of operation		6-36	36-67	67-83	83-102	102-122	122-141	141-154
HRT (h)		76	49	36	25	15	11	5.3
Influent Cephalexin (mg/L)		2.76	2.91	2.78	2.65	2.81	2.7	3.5
Effluent concentration (mg/L)	Cephalexin	0.561	0.542	0.539	0.536	0.534	0.841	1.08
	7-ADCA	1.25	1.358	1.411	1.464	1.466	1.141	1.521
	Acyl group	0.95	0.98	0.952	0.911	0.921	0.749	0.910
Removal efficiency (%)	Cephalexin	78.6	79.4	80.2	81.1	82.2	73.2	66.3

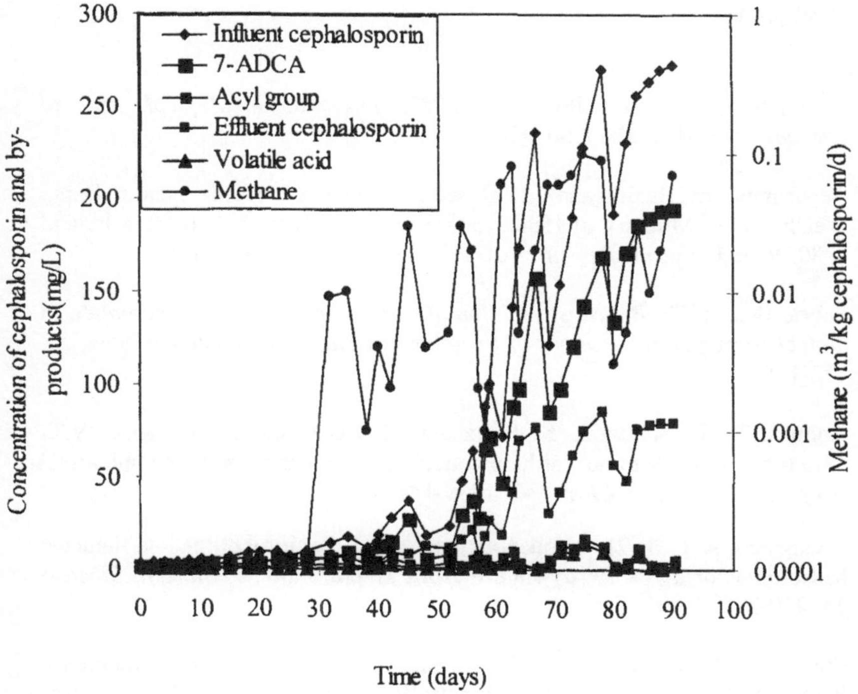

Figure 2. Effect of Influent Cephalosporin Concentration and its Transformation on Reactor Performance

CONCLUSION

The following conclusions were drawn from the results of this study: Cephalexin degrading cultures served as an excellent inoculum source for the rapid start-up of the anaerobic fluidized bed reactor, which was operated for the biotransformation of cephalexin. Continuous removal efficiency of 81% was obtained for a maximum cephalexin concentration of 175 mg/L at 3 h HRT using bioaugmentation through periodic addition of acclimated cells from an off-line enricher-reactor. Results of this study suggested that the anaerobic fluidized bed reactor with bioaugmentation was a feasible process for the anaerobic treatment of cephalexin based pharmaceutical wastewater.

REFERENCES

1. Goodman, L.S. and Gillman, A. 1980. *The pharmacological basis of Therapeutics*. Macmillan, pp. 95-110.

2. Department of Health and social security, Scottish Home Health Dept., Welsh office, Ministry of Health and Social Services for Northern Ireland 1980. *British Pharmacopoeia*, Vol. I, University Press, pp. 46- 65.

3. Sebek, O.K. 1977. *Biotransformation of steroid harmones and antibiotics, in biotechnological applications of Proteins and enzymes*. Academic Press, pp. 122-136.

4. Komori, T., K. Kunugita, K. Nakahara, H. Aoki, and H. Imanaka. 1978. "Production of 3-amino cardicinic acid from nocardicin C by microbial enzymes", *Agri. Biol. Chem.*, 42: 1439-1447.

5. Livingston, A. G. 1991. "Biodegradation of 3,4-dichloroaniline in a fluidized bed bioreactor and a steady-state biofilm kinetic model", *Biotech. Bioeng.*, 38: 260-272.

6. Ebert, C., L. Gardossi, and P. Linda. 1998. "Activity of Immobilised Penicillin Amidase in Toluene at Controlled Water Activity", *J. Mol. Catal. B: Enzymatic*, 5: 241-244.

7. De Martin, L., C. Ebert, G. Garau, L. Gardossi, and P. Linda. (1999). "Penicillin G amidase in low-water media: immobilisation and control of water activity by means of Celite rods", *J. Mol. Catal. B: Enzymatic*, 6: 437-445.

8. Babcock, R. W, K. S. Jr. Ro, C. C. Hsieh, and M. K. Stenstrom. 1992. "Development of an Off-line Enricher Reactor Process for Activated Sludge Degradation of Hazardous Wastes", *Water Environ. Res.*, 64 (6): 782-791.

9. Cardinal, L. C., and M. K. Stenstrom. 1991. "Enhanced Biodegradation of Polyaromatic Hydrocarbons in the Activated Sludge Process", *Research J. Water Environ.* Fed., 63: 950-957.

10. Ramanand K., and J. M. Sulflita. 1993. "Carboxylation and mineralization of m-cresol by a sulfate-reducing bacterial enrichment", *Current Microbiol.* 26: 327-332.

11. APHA-AWWA-WPCF. 1995. *Standard Methods for the Examination of Water and Wastewater.* 19th edition, APHA, Washington, USA.

AUTHOR INDEX

INDICES FOR 28TH, 29TH, 30TH, 31ST & 32ND INDUSTRIAL WASTE CONFERENCES